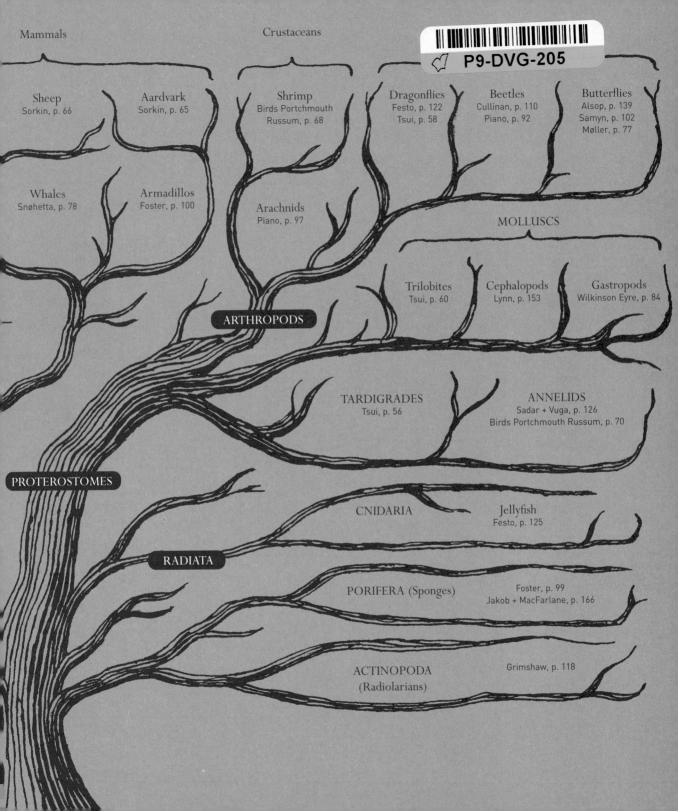

Mammals

Crustaceans

Sheep
Sorkin, p. 66

Aardvark
Sorkin, p. 65

Shrimp
Birds Portchmouth
Russum, p. 68

Dragonflies
Festo, p. 122
Tsui, p. 58

Beetles
Cullinan, p. 110
Piano, p. 92

Butterflies
Alsop, p. 139
Samyn, p. 102
Møller, p. 77

Whales
Snøhetta, p. 78

Armadillos
Foster, p. 100

Arachnids
Piano, p. 97

MOLLUSCS

Trilobites
Tsui, p. 60

Cephalopods
Lynn, p. 153

Gastropods
Wilkinson Eyre, p. 84

ARTHROPODS

TARDIGRADES
Tsui, p. 56

ANNELIDS
Sadar + Vuga, p. 126
Birds Portchmouth Russum, p. 70

PROTEROSTOMES

CNIDARIA

Jellyfish
Festo, p. 125

RADIATA

PORIFERA (Sponges)

Foster, p. 99
Jakob + MacFarlane, p. 166

ACTINOPODA
(Radiolarians)

Grimshaw, p. 118

zoomorphic

zoomorphic

NEW ANIMAL ARCHITECTURE

Hugh Aldersey-Williams

LAURENCE KING PUBLISHING
in association with
HARPER DESIGN INTERNATIONAL
an imprint of HarperCollins*Publishers*

LAURENCE KING

Published in 2003 by Laurence King Publishing Ltd
71 Great Russell Street
London WC1B 3BP
United Kingdom
Tel: + 44 20 7430 8850
Fax: + 44 20 7430 8880
e-mail: enquiries@laurenceking.co.uk
www.laurenceking.co.uk

Published in North and South America by:
Harper Design International
an imprint of HarperCollins*Publishers*
10 East 53rd Street
New York, NY 10022
Fax: +212 207 7654

A catalogue record for this book is available
from the British Library

ISBN 1 85669 340 6

Designed by Godfrey Design
Picture research by Peter Kent
Printed in Singapore

For news of the *Zoomorphic* exhibition at the
Victoria & Albert Museum and the author's
other projects, visit www.hughalderseywilliams.com

To Sam

acknowledgments

The collision of architecture and biology that takes place within the pages of this book has surprised many, perplexed some, and clearly pleased a good few of those with whom I have spoken during its preparation. It has been a joy to witness these reactions, and I have benefited as much from the probing criticism offered by some as from the instant camaraderie I found with others.

Zoomorphic relies for its success largely on the visual comparison of animals and buildings. It was conceived in the first place as an exhibition, and I am grateful to Susan McCormack and Shaun Cole at the Victoria & Albert Museum for taking it on.

The architects whose work appears in these pages supplied material and answered questions with great generosity, and have accepted without qualms the presentation of their work in a context beyond their control. I am indebted to them and their staff.

Many others have informed my awareness of the larger context in which this development is taking place. Among them are Thomas Arnold, Bernard Cache, Rosamund Diamond, John Frazer, Julian Gibb, Jan Kaplicky, Ken Yeang; Aran Chadwick, Brian Forster, Alan Jones, Jörg Schlaich, Jane Wernick; Mike Hansell, David Raup, Jim Smith, Colin Tudge, Julian Vincent. George Jeronimidis kindly inducted me into the work of the Centre for Biomimetics at Reading University, while Mandy Holloway of the Natural History Museum in London indulged and abetted my effort to find animal analogues.

I am grateful to all at Laurence King Publishing, especially to my editor Philip Cooper, who has provided both ideas and assistance beyond the call of duty, to Peter Kent, who rose to the challenge of some esoteric picture research, and to Robert Shore for his eagle-eyed editing. Jason Godfrey's design well captures the spirit of the movement I am struggling to describe.

Finally, it is a pleasure to thank Moira, my family, and Audrey Boughton (who is as good as) for their unwavering encouragement.

Hugh Aldersey-Williams
London, October 2002

contents

/*introduction*

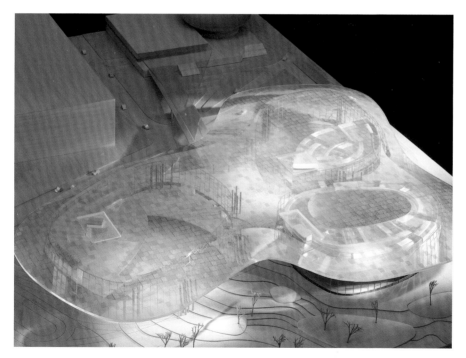

Zaha Hadid
**BMW competition
entry**
2001

Sauerbruch Hutton
**BMW competition
entry**
2001

In December 2001, the German car-maker BMW announced the results of a competition for the design of a customer reception centre at its headquarters and factory complex adjoining the site of the 1972 Olympic Games on the outskirts of Munich. The company chairman, Joachim Milberg, proclaimed his wish that the new building should represent 'the pioneering architectural core of the BMW brand'.

The firm of Coop Himmelb(l)au emerged triumphant from a pool of eight finalists. The Austrian architects' design, expected to be complete by the end of 2004, is highly flexible and transparent, not unusual qualities for flagship corporate buildings these days. On the other hand, its overall shape *is* unusual. The building is not a cool glass box, nor even a dramatically spiky shard. Instead, it gently bulges. The glass roof is conceived as two layers, one that swells upwards and one that bellies down, providing a ceiling of variable height to the spaces below, based on 'simulated reactions' to the way they will be used. Overall, the sinuous roof volume seems like the body of a breathing organism. Propped on Dalí-esque stilts, it looks like a blue whale washed up on the beach.

In fact, all the finalists' designs – and many of the 275 entries eliminated in earlier rounds – were characterized by irregular curves. The architects wrote of waves and forests, or inserted into their spaces forms said to resemble organs. Morphosis from California came fourth with a design in which 'dynamic, organic forms are allowed to generate themselves'; Britain's Zaha Hadid, in third place, described 'softly modulated spaces and forms'; runners-up Sauerbruch Hutton from Berlin squeezed three ellipsoidal pavilions between an undulating roof of clouds and an undulating valley floor with hardly a straight line in sight. For some, the chance to pay homage to Frei Otto's swooping Olympic stadium nearby – still looking novel after 30 years – was not to be missed. But this alone does not explain why so many leading architects

Morphosis
**BMW competition
entry**
2001

should be thinking similar thoughts.

This coincidence is remarkable enough. But in its immediate geographic context, it is still more startling. For the more obvious cue to which the architects might have responded was not Otto's stadium but BMW's own existing headquarters. So it is significant, especially considering the client, that they turned their back on the building next door. The 'four cylinders', as it is known, was completed at the same time as the Olympic architecture, its clustered aluminium towers 'a symbol of dynamism, innovation and technical progress', as BMW's literature proclaims. Today, it seems a garish celebration of the internal combustion engine, and in architectural terms represents a dubious apotheosis of the machine aesthetic.

This change in architectural fashion for one image-conscious client is a bellwether for a huge shift away from the mechanistic towards the biological in aesthetics and cultural rhetoric. Businesses no longer speak of re-engineering, but of adaptation and evolution, and everything from film to fashion to fine art is obsessed with life science. Paradoxically, the automobile industry is in the vanguard of this shift, using the full range of visual media to insinuate a connection between machine and nature. Advertisements for cars now routinely feature biological images, from dolphins to DNA, while in Britain the well-known geneticist Steve Jones has appeared on television to promote the 'genetic engineering' of the Renault Laguna.

There are two crucial reasons why the automobile industry should be among the first to move in this direction. First, it makes a kind of sense to indicate corporate concern for the environment – however slightly it is reflected in the actual products – by adopting nature-lover's language. Why architecture should emerge as one of the means to communicate this urgent message becomes more apparent when one considers the other reason why it is *this* industry which has taken the lead. For nearly 20 years now, car shapes have been changing

Kurt Schwanzer
BMW Headquarters
Munich, Germany
1972

in ways scarcely appreciated by the buying public. Computers have allowed designers to create more fluid shapes, fuelling a revolution in appearance that is now filtering through to other areas of product design – and belatedly to architecture. The signal moment was the launch of the Ford Taurus in 1986. 'I knew it was the future,' says Greg Lynn, one of the pioneers of computer-generated 'blobitecture', only slightly tongue-in-cheek.

In architecture, this development comes at an opportune moment. The old dogmas of both the Modernists and their repudiators have collapsed. Meanwhile, there are new materials and a new bravado among structural engineers that allow forms imagined on a computer screen actually to be constructed. The technical possibility and the cultural mood are in rare conjunction. Freed from the constraints – ideological and physical – that favoured rectilinear designs, architects are celebrating with an extravagant eruption of wild forms that go beyond the merely organic and promise to usher in a period of biological baroque.

This new architecture provides an emphatic answer to the critic Charles Jencks's recurrent question 'In what style shall we build?' There is the problem of what to call the style, however. 'Organic' has lost its precision, and tends to be applied loosely to anything with a few curves. Labels have been proposed such as 'biotechnic' or 'technorganic', but these imply a restrictive dependence of biological form upon technological means. Biomorphism, a term coined during the Art Nouveau period, remains more specific than 'organic', but suggests that it is only shape that matters, whereas it is also patterns and mechanisms of building use and operation derived from biological models that interest a number of architects today. Unfortunately, no one term comfortably encompasses the variety of the present trend.

This book and the exhibition that it accompanies focus specifically on animalistic (or zoomorphic) new architecture. Sometimes, animal forms

Ford Taurus
1986

are employed for symbolic and metaphoric reasons, sometimes because nature inspires an idea of structure or skin or building function. These projects provide the most astonishing examples of this new trend, and the most persuasive evidence of architecture's new turn to nature.

The fossil record of animal architecture

Some architectural critics dismiss the new biomorphic architecture as a joke in poor taste. But though it does betray an undeniable propensity towards kitsch at times, to write it off completely would be a mistake. There is an honourable and long-lived tradition of representation of natural form in human culture. In architecture, these flowerings have tended to be rather ephemeral. At the beginning of the 21st century, however, there is reason to think that a real breakthrough will finally be made. Not only is the technology now available to permit the affordable emulation of biological form, architecture's theorists can find no reason to forbid it, and the culture is receptive to it. Furthermore, we are in a position as never before to apply knowledge gained in the biological sciences to extend this development beyond mere style.

But this is a glimpse of a future scarcely imagined by most architects, whose knowledge of biology is negligible. So, instead, let's look back for

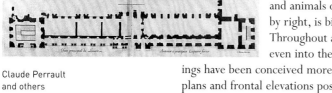

Claude Perrault
and others
Palais du Louvre
Paris, France
begun 1667

a moment. In doing so, we find that it is not entirely by chance that animals and buildings share some of their most basic characteristics. One trait that buildings and animals often share, as it were by right, is bilateral symmetry. Throughout architectural history, even into the 20th century, buildings have been conceived more often than not with plans and frontal elevations possessing a substantial degree of bilateral symmetry (i.e., a position on one side of a central axis can be reflected through that axis to an identical, but mirrored, position on the other side). This custom – for that's all it is; there's no *fundamental* reason why buildings should follow this rule – may be inspired by the bilateral symmetry of many animals, and of one in particular.

In the 16th century, Giorgio Vasari laid out his conceptual plan for an ideal palace on anthropomorphic lines – the façade was the analogue of the human face, the courtyard the body, the stairways the limbs. The only symmetry that such a design possesses is bilateral symmetry. In mathematical terms, this is a low-order symmetry. Higher orders of symmetry permit more numerous and varied mapping of the parts of a whole, not only by reflection but also by rotation and other means. Yet bilateral symmetry is special to us precisely because we recognize it from nature; we see it in every face we meet.

It is important to note that creatures do not *achieve* this state of pleasing symmetry; their development is a process of symmetry *loss*. Bilateral symmetry is all that's left after development from a spherically symmetric fertilized egg. Imagine this egg at the point of intersection of three lines, running up and down, left and right, and front and back. As the cells divide and divide again during the course of development, the organism loses symmetry. In some species, the point at which the sperm cell enters the egg cell may determine the initial polarity, but the mechanisms by which further polarization takes place are still poorly understood. Air-breathing animals must fight against the force of gravity; in the struggle, any up–down symmetry, that is symmetry about an imaginary horizontal plane perpendicular to the line of action of the gravitational force, is lost. (Many waterborne creatures effectively exist in a zero-gravity environment, and retain this symmetry in large measure.) The requirement to find food and escape predators demands a means of independent locomotion, which then destroys symmetry along the front–back axis. (Again, creatures that move up and down, such as

William Blake
The Tyger
1794

jellyfish, or that do not move at all, such as sea anemones, retain this symmetry.) That leaves only left–right symmetry, which presents no obstacle to survival, and so can stay.

In architecture, gravity is always a constraint. For this reason if no other, few buildings would look the same if turned upside down. But appearances about the vertical planes of symmetry, left–right and front–back, are determined by cultural rules. Buildings do not move, but people are expected to move through them according to certain social rules, and this leads to the standard requirement for a building to have a façade, where visitors are greeted and enter, and a rear, where waste is evacuated – a scheme that replicates animal functions. Then, as in nature, we are left with the bilateral symmetry surviving unchallenged – and we have arrived, by an unexpected route, at the 18th-century architectural theorist Jean-Nicolas-Louis Durand's conception of the principal and secondary axes of symmetry proper to architectural composition.

The fondness for bilaterally symmetric architecture has persisted from the Classical Greek period through the Gothic and the Renaissance, and has even weathered to a surprising degree the upheaval of the Modern Movement, as Colin Rowe showed in his famous essay *The Mathematics of the Ideal Villa*, where he pointed out the shared symmetry in villas by Le Corbusier and Palladio.

But bilateral symmetry is just a start. Perhaps, as Greg Lynn suggests, it is merely 'the cheapest form of beauty', nature's default setting. Certainly, symmetry of this or any other kind is neither a necessary nor a sufficient condition for a rich architecture or for a specifically biomorphic architecture.

Nevertheless, this unignorable parallelism between nature and architecture has encouraged some to extend the biological analogy. Nineteenth-century architects such as Eugène-Emmanuel Viollet-le-Duc and Gottfried Semper thought fit to adapt the anatomically based classification of animal species developed by the great French naturalist

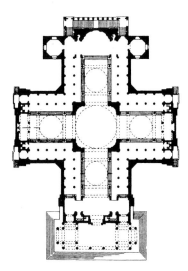

Jacques-Germain
Soufflot
**The Panthéon
(Ste Geneviève)**
Paris, France
begun 1757

Georges Cuvier to building types – although the exercise duly led them in different directions, favouring the Gothic and Classical styles respectively; neither architect was able to escape the influence of architectural history.

A little later, the steel frame pioneered by Louis Sullivan and the Chicago School from the 1880s meant that internal structure and external cladding could be regarded as separate. Comparison with skin and bone in vertebrates was inevitable. (In later, 'high-tech', architecture – a movement that is now producing many of the exponents of bio-morphism – the mass of a building is often borne by a structure external to the skin, which draws invertebrate phyla such as insects and marine arthropods into the analogy.)

But although the comparison was made, it was not pursued: Chicago's early skyscrapers did not look animalistic. It was another few years before there was an eruption of biological form unprecedented in architectural history. Although Art Nouveau threw up shapes that were predominantly vegetal – the Italians termed it *stile floreale* – there were also similarities with animals. The Jena zoologist Ernst Haeckel's *Kunstformen der Natur*, begun in 1899 at the height of the Art Nouveau and Vienna Secessionist movements, illustrates the resonance with its numerous obscure, often microscopic marine creatures drawn and coloured like so many Tiffany lamps. It is hard to believe that Haeckel's magnificent drawings of cilia, for example, did not directly inspire Bruno Taut's famous domed glass pavilion for the 1914 Werkbund exhibition in Cologne.

Art Nouveau was not the only impulse towards natural forms in architecture around the turn of the

Renzo Piano and Richard Rogers
Centre Georges Pompidou
Paris, France
1977

Victor Horta
Tassel House
Brussels, Belgium
1892

19th century. Rudolf Steiner's anthroposophy movement, a religion and philosophy based on the centrality of human development, achieved expression in architecture, notably at his school built in 1913, the Goetheanum at Dornach in Switzerland, where various elements of the building suggest human organs. The Hungarian architect Imre Makovecz is among those working today who have been inspired by this tradition. Both Art Nouveau and Steiner's theosophy left their mark on modern architecture through their influence on teachers at the Bauhaus and on architects such as Frank Lloyd Wright, who took 'organic' architecture to new heights.

If this was the sublime, then biomorphic architecture has also had its ridiculous side. For there is also a robust vernacular tradition of buildings that represent animals in the most literal way. Specimens such as the Big Duck – a shop in the shape of a duck selling duck decoys, celebrated by Robert Venturi and Denise Scott Brown – began to proliferate in the 20th century, although the trend was perhaps started by Lucy the Elephant, a seaside attraction in Margate, New Jersey, patented by one James Lafferty in 1883. Even here there are distinguished precedents, such as the byre designed to resemble a cow proposed by Jean-Jacques Lequeu in the 18th century or the Renaissance monster theme park of Bomarzo in the Lazio region of Italy in the 16th. A few artists and architects have continued to risk the label 'kitsch' by employing obvious animal forms in their work. The French sculptor Niki de Saint Phalle, wife and collaborator of Jean Tinguely, designed a number of habitable animal structures during the 1970s, while Frank Gehry netted his first architectural fish in the next decade.

Charles Jencks has claimed that 'Modern architects flirt with the concept and style of "organicism"… [e]very forty years or so'. But why only intermittently and, if intermittently, why the regularity? Over the (statistically insignificant) span of the 20th century, Jencks's approximation seems adequate based on the evidence. There *was* a lull in

Imre Makovecz
Stephaneum
Piliscsaba, Hungary
1999

interest following the demise of Art Nouveau and the First World War, and a revival during the 1950s, before the present resurgence beginning in the 1990s. The intervening periods of the 1920s–1930s and 1960s–1980s were dominated by the austerities of Modernism and Late Modernism.

The revival, if that's what it was, of organic architecture in the 1950s was made possible by Second World War developments in concrete construction as much as Art Nouveau was by the use of steel and glass. Architects and engineers such as Pier Luigi Nervi and Gio Ponti as well as Oscar Niemeyer exploited concrete's structural potential, while others pushed in a biomorphic direction. The triumphant projects of this decade are Eero Saarinen's aquiline TWA Terminal at New York's Kennedy Airport and Jørn Utzon's polysemous Sydney Opera House.

In fact, 'organicism' has been a fairly constant presence in architecture at least since Art Nouveau. Even when it has not been the mainstream fashion, there have been influential figures such as Wright to tend the flame. During the 1960s and 1970s, this duty was taken over by a troupe of counterculture provocateurs, among them Archigram in the United Kingdom, Habiter la Mer in France, Haus-Rucker-Co in Germany, and Ant Farm in the United States. Coop Himmelb(l)au's competition-winning project for BMW appears in a new light when compared with the firm's 'Cloud' projects from this period. As ever, new materials were crucial in shaping form — on this occasion, often flexible fabric-based structures held in place by tension cables or the pressure of inflation — and visionary engineers such as Buckminster Fuller and Frei Otto were able to bring some of the most ambitious schemes to fruition.

The catalogue of biomorphic architecture from the last half-century suggests that the caricature of

The Big Duck
Riverhead, New
York, USA
1931

Modernism's opposition to nature – reiterated recently in Alan Powers's *Nature in Design* and Susannah Hagan's *Taking Shape* – has been somewhat overdone. Despite the damage that the Modern Movement undoubtedly did to our sense of connection with nature, its greatest practitioners, men such as Alvar Aalto, Wright and Le Corbusier, all strove in their way to lay emphasis on this connection.

Hagan may have overstated the starting position, but she is surely right when she describes the contemporary trend: 'The oppositions between culture and nature, so importantly and brutally drawn up by modernism, are dissolving again, not in a return to what was, but a transformation of it... The division between the living organism and the machine continues to collapse.' More precisely, architecture's relationship to both of these other entities is in the process of being renegotiated.

Cuvier first found it instructive to construct an axis with the living organism at one end and the machine at the other. The triangle of organism–machine–architecture was then completed by Le Corbusier with his concept of the 'House-Machine', and by the many from Semper through to Frederick Kiesler who have favoured organic analogies in architecture.

BMW's new architecture, then, signifies not usurpation of the machine by nature but their reconciliation. Nature has not triumphed over the machine either in architectural metaphor or in fact. But as our knowledge increases of biological form and function, so new analogies and metaphors from living organisms commend themselves. At the same time, our comprehension of biological processes grows more detailed and 'mechanistic'. The building process too becomes more machine-like. The triangle is shrinking on all sides. The current resurgence of biomorphic architecture is not a reaction against

Pier Luigi Nervi
Exhibition Hall
Turin, Italy
1948

Modernism but its logical continuation, and for this reason may enjoy a more extended life than many critics are yet prepared to predict.

Science, function, metaphor and accident

While animal forms have always played a role adding some of the deepest layers of meaning in architecture, it is now becoming evident that a new strand of biomorphism is emerging where the meaning derives not from any specific representation but from a more general allusion to biological processes. The many new buildings that suspend blobs like amoebae within otherwise Cartesian volumes are illustrative of this new phenomenon; so are those computerized creations that use 'evolutionary' algorithms and quasi-genetic coding schemes as their preferred means of generating form.

Science has had metaphorical power for architecture ever since the two disciplines diverged in the 18th century. As Adrian Forty explains in Peter Galison and Emily Thompson's *The Architecture of Science*: 'Only when science became a field of knowledge separate from architecture would there have been any appeal in seeing architecture as if it was a science. Metaphors are experiments with the possible likenesses of unlike things. Each one of the countless scientific metaphors in twentieth-century architecture is a little experiment, an attempt to find a relationship between architecture and one or another branch of science, but they all rely on our belief that really, at the bottom, architectural practice is not scientific.' The majority of these little experiments have taken the physical sciences as their starting point – perhaps unsurprisingly, since these were the dominant sciences at least until the mid-20th century. Within this vocabulary, there has been a detectable shift of emphasis, from confidence in science, seen at its most exuberant in the crystalline

Oscar Niemeyer
São Francisco Church
Pampulha, Brazil
1943

19

Daniel Libeskind
Jewish Museum
Berlin, Germany
1999

buildings of the Expos of the 20th century, peaking with Montreal in 1967, to unease with it, as architects have belatedly attempted to assimilate the disturbing revelations of 20th-century physics.

Deconstructivism has provided the boldest use of physical science as architectural metaphor. Its jagged, misaligned forms are inspired by the uncomfortable truths of quantum theory and relativity, as well as the shifts and disruptions of everything from plate tectonics to Post-Structuralist literary theory. This architecture can strike a deep chord – as Daniel Libeskind's disorienting and shocking Jewish Museum in Berlin does – but in general it is weighed down by its intellectual baggage.

There is every sign that the new architecture based on biological metaphor will be more accessible. Biomorphism has profound connections with the history of architecture, but it also connects immediately with the public. 'More liberated and imaginative forms, unacceptable to major corporate clients a few years ago, are now actively sought,' as David Pearson points out in *New Organic Architecture*.

On one reading, biomorphism is merely a subset of the architecture stimulated by the sciences encompassed by what Jencks calls Complexity Theory (although, as with that other popular moniker 'chaos theory', there is no theory in the true scientific sense). And like Deconstructivism, it offers a formal language that has the potential to satisfy our hankering for cultural richness.

But there is a more pressing, and perhaps less noble, reason for building biomorphically: to signify commitment to the natural environment. It seems that a building that looks like a natural organism, or is generally softened in appearance, is apt to be considered more environmentally responsible (or responsive) than a conventional tower or box. This expectation was set in train in Britain by Ralph Erskine's Ark office building in west London (completed in 1991), which, although a commercial failure in its first few years of life, set a benchmark for

Ralph Erskine
Ark Office Building
London, England
1991

environmental standards, and signified the fact by means of its shape and, still more, by its name.

It is a natural enough urge to wish to display the good one has done, and we should not criticise too harshly. However, it is hardly essential that a 'green' building look naturalistic. The 'bioclimatic' skyscrapers of the Malaysian architect Ken Yeang, for example, look novel in skyscraper terms but they are not especially biomorphic. Nor, conversely, should we assume that all biomorphic buildings are paragons of 'green' design. Nevertheless, it is currently the case that buildings that have an environmental story to tell often tell it in this way. It seems likely that this is a transitional stage, and when at last every new building does more to minimize its environmental impact, this signalling will become unnecessary.

It is important to remain clear-sighted about the separate roles of science in architecture: as a practical aid where scientific knowledge gives rise to new building technologies; and as metaphor. Even when science is clearly understood to serve as a metaphor, it is easy to introduce a misplaced moral imperative. Jencks does it in *The Architecture of the Jumping Universe* when he asks: 'Why should one look to the new sciences for a lead? Partly because they are leading in a better direction – towards a more creative world view than that of Modernism – and partly because they are true.'

In fact, the 'complexitist' architecture stimulated by 'the new sciences' is just a style, or a set of styles, no better or worse than other styles. Any science may provide a source of inspiration to architects, which is as legitimate as any other source, whether linguistic philosophy (Deconstructivism), popular culture (Venturi's Las Vegas-ism) or architecture's own past (Post-Modern Classicism). For, as Adrian Forty warns: 'Many of the metaphors in the architectural lexicon come from science… but we should not assume that just because a term comes from science it will make a good metaphor.' Indeed, quantum mechanics and the 'complexity sciences'

Ken Yeang
Nagoya Tower
Nagoya, Japan
1997

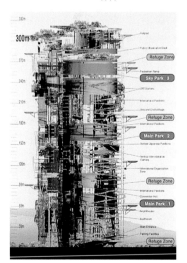

have nothing to contribute to architecture *except* possibly in terms of stylistic inspiration.

The biological sciences are no different in this respect. However, two factors argue for special consideration. The first is that biology provides a deeper well of metaphor which is tightly entwined with architectural history. The second is that, of all the sciences, it is biology that may now have most to offer in a practical sense for the development of new architectural technologies. But here it is important to recall Forty's distinction between architecture that *is* science and architecture that *references* science. For, if architecture is to capitalize materially on the lessons that biology offers, it will require something of a reinstatement of the pre-18th-century state of affairs when architects found it entirely natural to converse with scientists.

That architects have developed an interest, however vicarious, in biology is clear from the appearance of the many new buildings included in this book. It is helpful to divide these works into three groups: buildings that employ biological, and here in particular animal, form for symbolic purposes; buildings where the animalism emerges more or less logically from the functional programme; and buildings that appear to the observer to possess biological qualities even though it may never have been the architect's aim to incorporate them. These are only loose categories, and some of these projects could easily appear under more than one heading.

Animal symbolism is employed with different levels of ambiguity. The most obvious symbolism arises when a building is destined to serve as a beacon or gateway for its city. Santiago Calatrava's addition to the Milwaukee Art Museum is a prime example of the type. The animalism of the symbol gives the building, and by extension the place, a kind of personality. In other circumstances, the animalism serves as a metaphor for the building's purpose; transportation terminals are especially prone to this treatment. A still looser metaphorical approach is often seen in buildings with an environmental

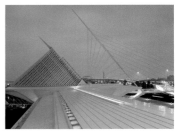

Calatrava Valls
Milwaukee Art Museum
Milwaukee, Wisconsin, USA
1994–2001

agenda, either, as at Nicholas Grimshaw and Partners' Eden Project, because of the nature of their contents, or because they make a claim to relative sustainability.

Modernism as an aesthetic – the International Style – may be in retreat, but the movement's underlying doctrine of functionalism remains unshakable for many architects, and a second important strain of animal architecture extends this functionalist tradition. For, if we believe – as we surely do, even if we haven't yet all the evidence to support the belief – that every part of a creature evolves to serve some function, then by following nature we seek to approach an ideal of total functionality. This is the root of the functionalist's wish to emulate nature; and the means to do so more exactly are fast becoming available, as computer-aided design and manufacturing allow Fordist economies of scale to be brought not only to buildings based, like Paxton's Crystal Palace, on the repetition of identical parts, but to edifices of more varied morphology assembled from unique components. Leaving aside those still detained by the aesthetics of repetition, the entire 'high-tech' school now finds itself logically positioned to draw new lessons and inspiration from biological form.

A central figure for this group is D'Arcy Thompson, the brilliant polymath who, in 1884, at the age of just 24, was made professor of biology at Dundee. His 1917 masterwork, *On Growth and Form*, set out to show that the shape of living things has a physical and mathematical basis, and thus that 'form is a diagram of forces'. Thompson's argument is completely general, applying to plants and to airborne, waterborne and land creatures of all sizes. He cites the Eiffel Tower and the Forth Rail Bridge – the sort of archetypes admired by the 'high-tech' fraternity – as sharing their form with the trunk of the oak and the skeleton of the bison respectively. The temptation is to conclude that man was inspired to follow nature's example. But Thompson's more perceptive point is that both man and nature take

Nicholas Grimshaw
and Partners
The Eden Project
Bodelva, Cornwall,
England
1996–2001

the most economical course of action prescribed by physical laws. Jørn Utzon and Louis Kahn recommended the book highly, and today architects as varied as Moshe Safdie and Marks Barfield continue to find it an inspiration.

The third kind of new biomorphic architecture is biomorphic by accident. Whether for reasons of structural logic or of dogma, rectilinear geometry has dominated the history of Western architecture. Now that this is no longer the case *de jure*, we see tentative efforts to establish a dialectic between the

Joseph Paxton
Crystal Palace
London, England
1851

regular and the irregular, between determined and indeterminate forms, between the grid and the blob, as Greg Lynn puts it. Richard Rogers's Bordeaux Law Courts and Will Alsop's Peckham Library, with their wooden vessel inserts, exemplify the trend, as do Wilkinson Eyre's Magna science centre, Rem Koolhaas's proposal for the French National Library, and Jakob + MacFarlane's restaurant at the Centre Pompidou, all of which place large blob forms in dominating orthogonal spaces, extending Kahn's conception of cylindrical *servant* and cuboid *served* spaces.

On the face of it, this development arises as a direct consequence of the widespread adoption of computer tools that allow irregular forms to be generated (and their structural integrity calculated) with greater ease. Yet we must ask why these forms so often lend themselves to *biological* interpretation. One reason may be our propensity, accustomed as we are to Cartesian coordinates, to read any other geometry as organic. But a more significant reason is surely that biology has become a major ingredient in the contemporary cultural discourse, and that architects are designing biomorphically, and people are reading biomorphism into their work, even when this is no conscious part of the architects' agenda.

Some architects are determined to push the role of the computer even further. They see a chance to liberate architecture not only from the old formal rules but also from the creator's ego. 'The emerging architectural culture is less to do with the architectural mastermind and much more to do with the tools that we use,' asserts Tom Verebes of Ocean D, one of the firms at the forefront of this movement. Typically, the computer is given a number of parameters and set running to generate architectural form through a sequence of iterative calculations. The suggestion is that this is an 'evolutionary' process, and some proponents of it have even based their algorithms on biological data such as DNA or gene sequences. There are, however, serious flaws in this analogy, and it is a relief to learn that these architects in fact often cheat by intervening, God-like, in the evolution to pick out some especially pleasing mutation.

Modernism with legs

Whatever the motivation of the architects responsible for them, the buildings shown here, none much more than a decade old, exemplify a trend that is global, and as it were biodiverse, running across many phyla of the animal kingdom. My focus on animal architecture may seem capricious, and perhaps requires a little further explanation. I have chosen to make several exclusions. First, the plant kingdom. Any discussion of organic or biomorphic architecture should clearly include plants; however, plant forms in architecture are so often reduced to mere ornament that to include them here might be counterproductive. Although man is an animal, I have also opted to omit most anthropomorphic architecture, as discussion of this would open up areas of human and sexual psychology beyond the scope of this work. Finally, I omit architecture inspired not by the form of animals but by the structures that animals themselves construct. These architectural analogues have been more than adequately covered in the works of Bernard Rudofsky, Mike Hansell and Karl von Frisch.

I have favoured architecture where the whole building has a kind of equivalence with a whole organism because it offers the most persuasive evidence of biomorphism; there are very many more buildings where individual details have animal resemblances. Even in some of these 'obvious' cases, it would be natural to react by observing that they don't look all that much like animals. But 'a kind of equivalence' is to be preferred – a building that looked exactly like a particular animal would soon pall. Good architecture is richly allusive, and these buildings fascinate because they can be read in many ways. Le Corbusier's Ronchamp chapel suggests a bird metaphor, but also a nun's wimple, a boat, an open hand and more (as was revealed by Hillel Schocken in a famous seminar at the Architectural Association in London); it may even be read as a chapel, a conventional building type. Likewise, these works of architecture have many meanings, some doubtless intended by their architects, others certainly not – for we are all entitled to interpret a building as we may.

Richard Rogers
Law Courts
Bordeaux, France
1998

Is the present eruption of biomorphism any more than a flash in the pan? Will it persist any longer than Art Nouveau? Is biological form of the kind that architects are now able to realize in any way truly apt for our lives? Does it reconnect us with nature in any meaningful way? Is it form that humans feel specially comfortable with? It is perhaps too early to answer these questions. The morphology is clearly revolutionary, and perhaps in due course the underlying architectonics will be revolutionized as well. (At the moment many bio-morphic buildings are necessarily but illogically constructed on straight steel or wooden frames, just as the floating Modernist stucco of Le Corbusier's villas hid more prosaic structure behind.)

Talk of movements and schools and the sticking-on of architectural labels now seem rather passé, but is this a new modernism? It is not *the* new modernism; it will remain one approach and style among many. It is, rather, a significant, fertile and authentic manifestation of a continued and revivified Modernism, which has the potential, like 20th-century Modernism, to excite and connect with the public and to function in more intuitive and less hierarchical ways.

In his influential 1922 collection of essays, *Form in Civilization*, the influential English architect William Lethaby called for walls and vaults to be catalogued like the parts of the biological cell. 'Some day,' he wrote, 'we shall get a morphology of the art by some architectural Linnaeus or Darwin, who will start from the simple cell and relate it to the most complex structure.' No need for this now! The components of architecture turn out to be not so fundamental as Lethaby supposed, and certainly less permanent than their biological analogues, and there is growing recognition that biology has deeper lessons to offer.

If these lessons are assimilated, there is reason to believe that biomorphism will provide a rich and durable theme in 21st-century architecture. It is, as we have already seen, popular with the public, and above all there are the means to realize it. When Henry-Russell Hitchcock wrote his catalogue essay for the New York Museum of Modern Art's exhibition of Art Nouveau in 1960, he was obliged to note 'the excessive amount of invention that even a modest Art Nouveau structure required'. Hector Guimard's Paris Métro stations, he observed, achieved certain economies of scale because they were mass-produced, but individual houses never did. Art Nouveau faded because it required too much design time, expensive skills and custom components (and the same might be said for the concrete expressionism of the 1950s). Today, computers promise to alleviate these difficulties.

So what are we letting ourselves in for? Not all biologically inspired architecture will be good, but some of it will be the most astonishing architecture ever seen. Lethaby worried about the likely floridity of architecture inspired by nature, but consoled himself that it couldn't be any worse than the Baroque. Today, Jim Eyre of Wilkinson Eyre extends this thought: 'What would the architects of the Baroque have done given computers?' We can surmise that today's computer-assisted biomorphism is tame stuff yet. The visual analogy is frequently only expressed at the largest scale, and the focus so far has been on the *superficial appearance* of an organism evoked by the integument of an individual building, with little attention given to the biological lessons that might apply to the building's physical and environmental *performance*.

So far, many of these buildings have been civic trophies – museums and galleries rather than houses, offices and apartment blocks. We are invited to regard them as we regard art – as individual works set down in a neutral context that throws their shape into relief like sculpture in a city plaza. But these are buildings, not sculptures, and we are obliged to give them more than this passing consideration. What about the interiors? A building that looks like a starfish or a sea anemone is all very well, but what would it feel like to live inside one? And what about their relation to other buildings? How many buildings like this can a city stand?

Present objections may be based on little more than taste. After all, it's not often that we are confronted with such novelty in architecture. Similar questions were asked in the early days of Modernism, before it was found that its buildings could provide an enjoyable user experience and participate in the urban landscape.

With the help of the few examples we have so far, we can try to anticipate how the new biomorphic architecture will be assimilated. The radical exteriors may in fact hold the key. For unlike many conventional buildings, the external form of an animalistic building induces particular expectations of the space within. One cannot tell that a Schinkel façade conceals a magnificent rotunda or that a Le Corbusier villa contains a double-height living area, but one can be fairly sure that some organicity will continue inside a building with a biomorphic exterior. What will it feel like to inhabit and move through these spaces? Are all rooms to become wombs, all corridors colons and capillaries? Our understanding of traditional architectural space based on the 'rules' of proportion is learnt; is our appreciation of this alternative innate? Will it feel natural, like a home-coming? Or will it feel alien? Will we feel trapped, like Jonah in the belly of the whale?

Some organic forms may indicate how they are to be experienced – an entrance that is clearly an orifice may no longer need to be codified as a 'door-way'. Kathryn Findlay of Ushida Findlay uses the analogy of a worm eating its way through an apple to explain the development of spaces within her buildings. The movement of users through a building helps to determine the size and sequence of internal spaces and their overall shape. The interiors are 'reactive' to the body and perhaps feel more natural or comfortable as a result.

At present, the prime examples of the new biomorphism effectively stand alone in a landscape or in conscious juxtaposition with rectilinear sur-roundings. How, when the time comes, should one biomorphic building relate to another? What are the chances of a successful biomorphic urbanism? The natural world (obviously) experiences no conceptual difficulty in juxtaposing organic forms, which may be similar or wildly contrasting. Evolution simply makes it happen. Competition in nature offers 'design' solutions that may be characterized by relative order, as when certain trees in arid environ-ments position themselves at regular intervals on a grid for optimal water uptake, or relative chaos, as in a tropical forest where highly varied plant species compete in different ways for light and water. In architecture, then, biomorphism may beget

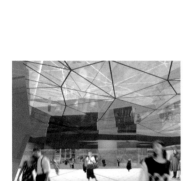

Alsop Architects
**Mauritsweg
Redevelopment**
Rotterdam,
The Netherlands
2001–

biomorphism, but this need not lead to a narrow definition of the style or to excessive sameness. All this promises an exciting contrast with the narrowed expectations produced by International Style Modernism.

A connection begs to be made here. In his *New Theory of Urban Design*, Christopher Alexander writes: 'When we look at the most beautiful towns and cities of the past, we are always impressed by a feeling that they are somehow organic. This feeling of "organicness," is not a vague feeling of relationship with biological forms. It is not an analogy. It is instead, an accurate vision of a specific structural quality which these old towns had… and have.'

There is an element of teleological circularity in this argument – do we love these old towns because of their organic quality, or do we perceive an organic quality in our effort to explain why we love them? The experience has nevertheless inspired Alexander to enumerate various rules for devising new urban plans. Some rules are close to nature – for example, the idea that urban structures should be allowed to develop without a plan, or the preferment of bilateral symmetry for important spaces. But other rules negate the Darwinian model, especially one that privileges 'Vision'.

These principles are largely untested, although Zaha Hadid's masterplan for Singapore and a 1993 proposal for the coastal Chinese city of Changhiu by Jeffrey Kipnis and Tom Verebes at the Architectural Association show how patterns from nature may influence planned cities in future. Alexander's model is not of course predicated on buildings themselves possessing biomorphic qualities, and the opportunity has not yet presented itself to bring biomorphic

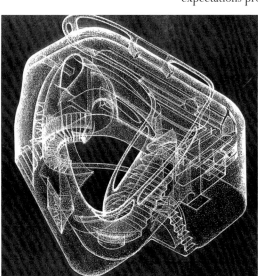

Ushida Findlay
Truss Wall House
Tokyo, Japan
1990–1993

buildings and organic urban planning together on any scale.

Nature itself doubtless contains many fascinating answers to both these conundrums – how to plan biomorphic interior space, and how to pack biomorphic buildings in urban space. But why should we pay particular attention to them? The reason why nature's answers might be worth more than other answers is that nature has to be economical. Each species has developed over millions of years so that it does what it needs to do with the minimum means. It is clear that taking natural forms as models can lead to lighter structures and more efficient use of materials as well as novelty of style, provided that the model is followed through appropriately. With greater dialogue between the scientific and architectural disciplines will come the prospect of other improvements – in thermal performance, weather protection, sensory responsiveness and so on – that are the subject of investigation in the emerging field of 'biomimetics'.

There is no *moral* imperative for this new architecture. Just because it seeks to follow nature does not mean that it is 'right' or pre-ordained. But as Aristotle observed, if there is a better answer to a problem, then nature has probably already found it. It is now up to architects and biologists, technologists and engineers to find those answers too.

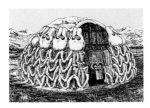

/*timeline*

architecture

1438
The Inca capital Cuzco, Peru, is replanned in the shape of a puma, with the fortress-temple of Sacsayhuamán at the head

8000 BC+
Palaeolithic dwellings use animal skins and sometimes bones for structure. A shelter uncovered in the Ukraine was found to comprise 385 mammoth bones

c. 2500 BC
Sphinx, Giza, Egypt

1500s
Renaissance monster park, Bomarzo, Lazio, Italy

1680s
The spire of the Church of Our Saviour, Copenhagen, resembles a spiral shell

1.75m years BC
Earliest known manmade shelter at Olduvai Gorge, Kenya

30,000 BC
Earliest known cave art, southwest France

8500 BC
Saharan rock art depicting animals

1200s
Gothic style, gargoyles

biology

1758
Carl Linné, or Linnaeus, makes zoological additions to his *Systema Naturae* using the system of binomial nomenclature for the classification of species that survives to this day

400s–1100s
Medieval bestiaries

1674
Leeuwenhoek observes many micro-organisms for the first time using optical equipment of his own invention, bringing knowledge of the great diversity of life

c. 70 AD
Pliny the Elder's *Natural History*

1749+
Buffon's *Histoire Naturelle* presents many animal species, specimens kept in the Jardin du Roi in Paris

9000 BC
First domesticated animals

c. 340 BC
Aristotle's *Historia Animalium*

1740s
Amateur naturalism flourishes

1770s
Structural iron-
work expands; the
British tend to
favour cast iron,
the French
wrought iron

1850s
Frederick Law
Olmsted, urban
parks

1854
Viollet-le-Duc
begins to publish
theoretical works
which revive Gothic
architectural ideals
using steel and
other modern
materials

1800s
The town plan of
Musumba in Congo
takes the form of a
turtle, possibly
because of the
defensive function
of its shell

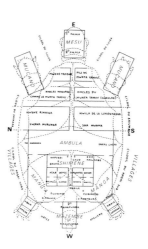

1855
Mild steel begins
to be made using
Bessemer's
process

late 1700s
Jean-Jacques
Lequeu designs a
dairy in the form of
a cow

1805
Jean-Nicolas-
Louis Durand
catalogues build-
ing form, type
and function

1812
Comparative
anatomical studies
lead Cuvier to
group the animal
kingdom into four
prototypical phyla,
destroying previous
notions of a linear
'chain of being'

1828
Baer's studies refute
the idea that the
development of an
embryo of a species
is in any way a sped-
up version of evolu-
tion of that species
– ontogeny *does
not* recapitulate
phylogeny

1850s
Thoreau celebrates
the return to
nature

1860s
Morphology is
used as the key to
evolutionary history
until mounting
palaeontological
evidence proves
more informative

1859
Darwin publishes
*On the Origin of
Species*

1820s+
Western zoos and
botanical gardens
flourish, bringing
many exotic
species to the
public gaze for
the first time

1827–38
Audubon's *Birds of
America*

1849
Kosmos presents
Humboldt's vision
of a globally
connected environ-
ment, the summa-
tion of his expedi-
tions begun 50
years previously

1860s
Herbert Spencer
asserts that
society can be
regarded analo-
gously to a biologi-
cal organism, and
draws parallels
between social and
biological evolution

1824
Dinosaur remains
first uncovered

1883
Lucy the Elephant at Margate, New Jersey, a patented design by James Lafferty, is the first of many American vernacular structures in the shape of animals

1884
Gaudí, already designing the extravagant buildings of the Park Güell, receives the commission for his life's work, the church of the Sagrada Familia

1895
Bing opens the first Salon de l'Art Nouveau in Paris

1899–1904
Guimard, Paris Métro entrances

1909
Berlage's St Hubertus hunting lodge near Arnhem has a plan based on deer antlers

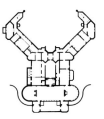

1910
First air-supported structures, F.W. Lanchester

1912
Pavel Janak, pioneer of Czech cubism, employs zoomorphic forms

1913
Rudolf Steiner builds the Goetheanum at Dornach, Switzerland, to promote his theory of anthroposophy; building elements suggest human organs

1880s
Early efforts at conservation; first nature reserves

1876
HMS *Challenger* returns from voyages of exploration with many marine specimens never before seen

1893
The term 'ecology', originally coined by Haeckel in 1866, is taken up at scientific conferences

1900
De Vries lays the basis for modern gene theory, relying in part on renewed interest in Mendel's neglected work on hybridization

1904
Haeckel at Jena publishes *Kunstformen der Natur*, beautifully illustrating many unfamiliar species, especially marine fauna

1914
Bruno Taut designs the Glass Pavilion for the Werkbund exhibition at Cologne

1940
Engineer Robert le Ricolais promotes the use of space frames in architecture and explores lightweight structures inspired by zoology

1943
Wright begins work on the Guggenheim Museum, New York

1950–55
Le Corbusier, Ronchamp Chapel

1921
Mendelsohn's Einstein Tower, Potsdam, Germany

1930s
American 'billboard architecture' makes abundant use of animal motifs

1939
Frank Lloyd Wright, Johnson Wax building, Racine, Wisconsin

1947
Aalto's Baker House dormitory at the Massachusetts Institute of Technology is dubbed by students the 'pregnant worm'

1948
Pier Luigi Nervi's Turin Exhibition Hall; along with Ove Arup and Felix Candela, Nervi represents a new breed of engineer interested in expressive form

1950s
Frederick Kiesler's 'house as organism'; architects and engineers explore the expressionist possibilities of concrete technology improved during the Second World War

1920s
Eileen Gray's circular studio

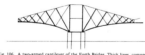

Fig. 106. A two-armed cantilever of the Forth Bridge. Thick lines, compression-members (bones); thin lines, tension-members (ligaments).

1953
Watson and Crick elucidate the structure of DNA

1917
D'Arcy Thompson publishes *On Growth and Form*, memorably comparing the Forth Rail Bridge with the skeleton of a bison, and stating that form, whether natural or manmade, is 'a diagram of forces'

Fig. 104. Skeleton of a fossil bison. From O. P. Hay, Iowa Geological Survey Annual Report, 1912.

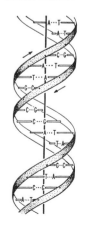

1953–61
Bulls' horns are among the visual devices encoded in Le Corbusier's government buildings at Chandigarh

1960s
Ant Farm, Sausalito, California, design 'The World's Largest Snake'

1962
Eero Saarinen, TWA Terminal, New York

1960+
Buckminster Fuller, geodesic domes

1961
Herb Greene, Prairie House, Norman, Oklahoma

1960s
Proliferation of tensile and lightweight structures

1957
Utzon, Sydney Opera House begun

1960s
Playful pneumatic structures proliferate: Haus-Rucker-Co's Pneumacosm cascades bubble-like dwelling units down a skyscraper, while Yellow Heart is a pneumatic love-nest; Coop Himmelb(l)au's 'Cloud' projects

1958
Eero Saarinen, Ingalls hockey rink, New Haven, Connecticut

1960
Lucio Costa's city plan of Brasília resembles a bird in flight

1964
Kenzo Tange, Tokyo Olympic stadiums

1962
Rachel Carson's *Silent Spring* catalogues humankind's depredations of the environment

1975
Foster Associates' Willis, Faber and Dumas offices, Ipswich, have an amoeba-like plan and a turf roof

1973+
Habiter la Mer, sea surface and undersea projects

1989
John Frazer and the Architectural Association's Diploma Unit 11 investigate the 'evolutionary' generation of form

1968
Roger Vadim's cult film *Barbarella* provides key images for architects of soft form

1972
Niki de Saint Phalle, Nellens house, Knokke, Belgium

1972
Frei Otto's Munich Olympic Stadium shows the potential of tensile membrane structures long before there are computers available to do the structural calculations

1984
Future Systems' projects begin to adopt more animalistic forms

1981
Gehry's fish motif first appears

1980s
CAD use increases rapidly

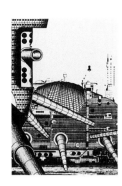

1964
Archigram: Ron Herron's Walking City and David Greene's Living Pod are among the group's animalistic projects

1967
Nicholas Grimshaw's first built project is a helical addition to the Anglican International Students' Club in London. Washrooms are easily accessed off a spiral staircase arranged not unlike the functional chemical groups in helical molecules in biology

1965
Hugh Casson, Elephant house, London Zoo

1972
James Lovelock publishes papers outlining the Gaia hypothesis that the earth is a system that regulates itself in order to maintain life

1976
Richard Dawkins, *The Selfish Gene*

1975
Edward O. Wilson publishes *Sociobiology*, proposing that the social instinct is the product of evolutionary adaptation, and so providing a new rationale for those who wish to make architecture more biological

1972
The Club of Rome publishes *Limits to Growth*

1966
Palaeontologist David Raup uses a computer to generate a rich diversity of shell forms

1973
Oil crisis; environmental consciousness increases

1989
'Green consumer' movement

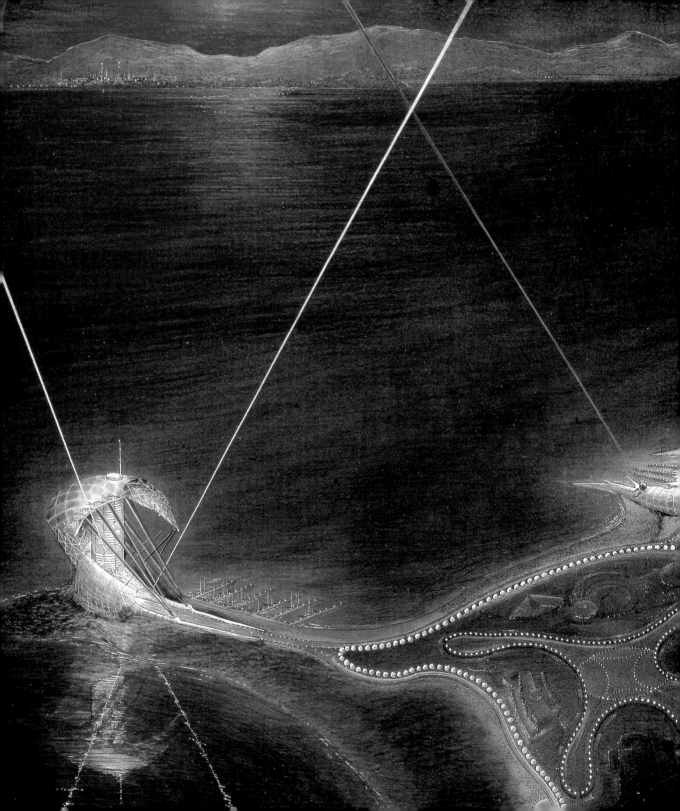

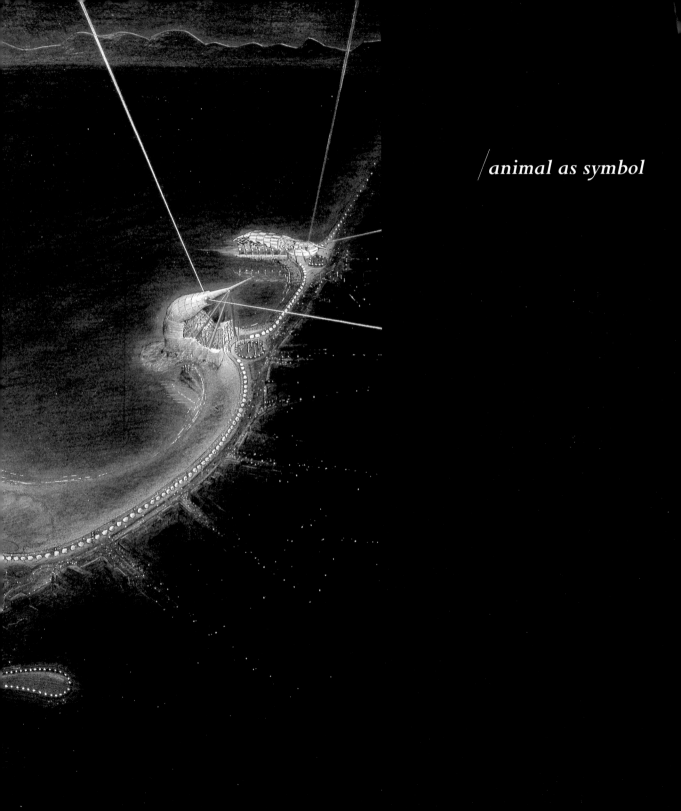

animal as symbol

Tsui Design and
Research
**Florence and
William Tsui
House**
Berkeley,
California, USA
1993–1995

From the ox and the ass at the Christian nativity to
the bull and bear of the financial markets, animals
are a more or less familiar means of representing
abstract ideas. A century or two ago classical painting was more deeply comprehended because people
knew that the animals they depicted represented
episodes in the Bible, myth or legend. Animals were
used to signify the Four Elements and the Four
Quarters of the World as well as assorted fates,
senses, sins and virtues. The butterfly emerging from
its chrysalis represents the soul departing the dead
body. The ape represents the devil or fallen man. In
a sign system laden with ambiguity and redundancy,
some species acquired a bewildering catalogue of
attributes, such as the scorpion, which signifies
hatred, envy, Africa, earth or logic according to
context, or the dove, indicative of lust and chastity
and the Holy Ghost.

As for painting, so for building. According to
Paolo Portoghesi: 'The symbolism attributed to animal
bodies by numerous civilisations has made it possible
for the architect to use symbolic imitation to communicate ideas and confirm collective values.' But
what does it mean today that a dolphin might allude
to the great fish that swallowed Jonah in the biblical
allegory of Christian faith, or that the lizard is a representation of Logic, one of the Seven Liberal Arts?

We are delighted, perhaps reassured, though
somehow not surprised when we learn that an
Australian aboriginal cultural centre incorporates in
its design elements of animal symbolism. When
there is a 'primitive' or religious agenda, we accept,
and perhaps even expect, this kind of symbolism.
However, all buildings are symbols – of power, of
wealth, of learning, of many things. How these
meanings are encoded is subject to change, but
throughout architectural history animal forms have
played their part, and we must assume that animal
form when it appears in contemporary building is
significant too.

Meanings are perhaps less singular than in the
past. In Jungian psychology, the Self is often repre-

sented by an animal, although the particular species seems immaterial; it may be a general mnemonic for our immersion in nature. Just as a miscellany of animals in a painting might be meant to remind the viewer of the story of Noah's Ark, or Adam naming the animals, or simply the Creation, we might today update the interpretation of the symbol to refer to global biodiversity. (Historically, artists have favoured the higher classes of animals, the mammals and birds, as symbols, whereas present-day examples of animal architecture range across all the major phyla of the animal kingdom.)

Significance may attach to animals at the level of their zoological class or order. Birds and fish, for example, adept in media we humans cannot master, have always symbolized transcendence. The fact that a bird is a classical representation of the ancient element of air surely adds a nuance to its frequent appearance in airport architecture. The representation of marine species in various seaside buildings seems equally to indicate a certain freedom from the usual architectural rules.

Animal form may reflect a building's importance as a civic symbol, or it may be the personal obsession of its architect – who, because of the dominance of rational Modernism and our concomitant embarrassment when confronted by symbolism, is liable to be branded an eccentric or a genius. In either case, the Jungian view seems to apply well enough, that the abundance of animal symbolism in the art of all periods indicates a wish to embrace into our lives not the extrinsic nature of each beast but their shared intrinsic quality – the instinctual.

Calatrava Valls
**Lyon Airport
Station**
Satolas, Lyon,
France
1989–1994

Gehry Partners

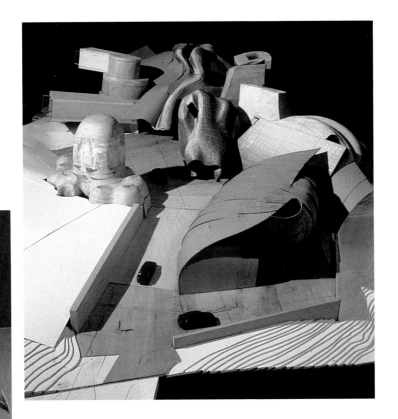

Lewis House (with Philip Johnson Architects) Lyndhurst, Ohio, USA 1989–1995 (unbuilt)

The large Lewis House with its requirement for various set pieces – a restaurant-standard kitchen, an indoor lap pool, guest houses and so on in addition to the usual facilities – called for the boldest sculptural treatment. Gehry collaborated with the architect Philip Johnson and artists including Richard Serra and Claes Oldenburg in devising a bewildering array of buildings dominated, in the early planning stage, by a massive, black, whale-like form

Smith House
Los Angeles,
California, USA
1981 (unbuilt)

Gehry uses a broad
range of materials
found in American
vernacular archi-
tecture to unac-
customed effect.
The cruciform
kitchen is green-
tiled, the bath-
house is brick and
the service 'shed'
is metal-clad.
Fibreglass and ply-
wood are among
other low-grade
materials employed
in a high-class
way. A colonnade
of eagle and fish
objects was
intended to greet
visitors coming up
the driveway of the
Smith House,
preparing them for
the fact that build-
ings of the new
house itself
resemble sculp-
tural objects

Frank Gehry is *sui generis*. His influences lie more in sculpture than architecture. His materials are drawn from the North American vernacular. Handled with great invention and dexterity, they become the ingredients of a cohesive and convincing personal style. Even flashy materials like the titanium used to clad the Guggenheim Museum in Bilbao are handled in a quasi-vernacular way — and so skilfully that the result is, according to Charles Jencks, 'a new well-scaled grammar of curving surface that can overlap, like fish-scales or the hide of an armadillo'.

Gehry has taken Modernist principles and subjected them to his own formal games. Internal function generates external form in unlikely ways: the stairwells in buildings for the furniture manufacturer Vitra near Basel, for example, become worm-like rectangular extrusions. The rooms of his houses are given distinctive forms and finishes.

This personal development would have been seen to great effect at the Smith House in Los Angeles. A new owner had commissioned Gehry to design an addition to a house he had done early in his career. The new house broke down the hierarchical plan of the old into fragmentary 'objects' with their own character. The driveway to the house would have been lined with more 'objects' — sculptural eagles and fish. In the event, the Bel Air Fine Arts Commission vetoed the project, saying it didn't look enough like a house.

The docks at Kobe proved less sensitive. Here Gehry's fondness for sculptural form came triumphantly to life in the design of a restaurant. The building houses various bars and dining areas in a snake-like copper-clad spiral and an adjoining shed-like volume. Sailing above it all is a giant fish sculpture made of chain-link mesh.

The fish has become a mascot for Gehry, but has also served as a heuristic from the days of his early collaboration with the sculptor Richard Serra around the time of the Smith House. It has reappeared in his temporary offices for Chiat/Day, and at the 1992 Olympic village in Barcelona, and reached its apotheosis with the proposed Lewis House in Lyndhurst, Ohio.

Why fish? The device began as a mocking response to Post-Modern Classical architecture at the beginning of the 1980s. But it is an all-purpose beast, and at Barcelona it seems instead an assertion of stylistic freedom in contradistinction to the ultra-Modernism of Skidmore, Owings and Merrill in the person of Bruce Graham who did the main buildings.

In both cases, the answer to the question 'why *fish?*' is 'why not?' And yet a fish specifically acquires extra significance because it is recurrent in architecture, even within the Modernist canon, appearing for example in Kenzo Tange's 1964 Olympic stadiums and Eero Saarinen's hockey rink at Yale University. For Gehry too it is a personal allusion to the childhood memory of his grandmother's habit of keeping the carp destined for the Sabbath dinner in the bathtub.

But he is still uneasy about it: at Kobe, the client made the suggestion, though aware of Gehry's preoccupation. And at Barcelona, Graham talked him into it. Gehry had been fiddling with other formal devices before he tried a fish. 'Almost immediately this idea worked better from all aspects,' he says. 'It not only complemented and strengthened the tower and its presence, but it also created a much stronger public space.' Perhaps he protests too much. For as Fiona Ragheb observes in her recent book on Gehry, the frequent appearance of the fish is 'testament to the functional appeal of the form's structural flexibility'.

The slippery form of the fish served as an exercise piece as Gehry's office made the transition to computer-aided design. Since developing the Barcelona fish, the office uses a computer programme (CATIA) adapted from the French military aircraft manufacturer Dassault Systèmes, but technology is not the driver of the architecture – 'the imagery on the computer takes the juice out of an idea,' he once remarked, and since his ideas have lots of juice, we should believe it. Gehry himself still starts his design process with fluid sketches and exploratory maquettes. These prototypes are then scanned and converted into digital surface models. The technology is then used to engineer the form so that it will stand up.

But this does not mean that the form is necessarily logical in conventional structural engineering terms. Gehry is not especially interested in engineering rigour or the honesty of transparent structure like many Anglo-American technophile architects. The skins of his buildings are said by these purists to conceal all manner of heavy-handed details. But the software has enabled Gehry to conceive progressively more elaborate sculptural forms.

This seems to be leading the architect in new directions away from the obvious figuration of the fish buildings. The major tendency is towards greater abstraction with heightened expressive power. The new Disney Concert Hall in Los Angeles is the heir to Bilbao in this regard. But the contemporary DZ Bank building represents a severe alternative, the consequence of a sensitive site near the Brandenburg Gate in Berlin and perhaps of collaboration with Jörg Schlaich, one of today's leading engineers of tensile structures, who cut his teeth working with Frei Otto on the Munich Olympic stadium.

DZ Bank Building
Berlin, Germany
1995–2001

Jörg Schlaich was one of the team of structural engineers on the oyster-like curved glass atrium roof and floor at the DZ Bank building in Berlin. The form has a mathematical purity missing in most of Gehry's work, but there is a residual suggestion of his trademark fish in the diamond-shaped panels of the swelling surfaces. The DZ building includes residential apartments as well as the bank's head office arranged around a large atrium. The steel-clad conference hall occupies the middle of the atrium like the pearl in the oyster

Milwaukee Art Museum
Milwaukee, Wisconsin, USA
1994–2001

Located on the axis of Wisconsin Avenue, the main street of Milwaukee, and connected to it via a cable-stayed footbridge, Calatrava Valls's pavilion makes a dramatic statement when seen either from Lake Michigan or the city. Symmetrically cantilevered from either side of a tilted spinal hinge, 72 slender beams, between 8 and 32 metres long and made of stiffened welded steel plates, are raised and lowered at different rates to form a huge kinetic sculpture. The physical resemblance to the bones of a bird's wing is actually rather slight, but as a symbol it is no less effective for that

Santiago Calatrava gives everybody problems – everybody, that is, except for the public, who instantly comprehend his buildings. Architects sneer that he is an engineer, while engineers find fault with his structural logic and dismiss him as a sculptor.

One strand of Calatrava's formal inspiration clearly stems from Art Nouveau (and the Gothic) and perhaps especially from his compatriot Antoni Gaudí. But contributing equally to his work is a rigorous engineering tradition from Switzerland, the country he has made his base. Here Robert Maillart and Christian Menn stretched the structural and expressive power of concrete in designs for bridges that had both to span challenging gorges and to fit in with beautiful landscape. Such constraints frequently lead to elegant, tapering forms that begin to suggest limbs and branches.

Calatrava's structures are not always minimum engineering solutions; the sculptor in him sometimes makes them heavier or changes the balance of the parts. This may offend literal-minded engineers, but it magnifies the expressive potential, and has led to Calatrava's becoming perhaps the pre-eminent contemporary architect of symbols.

His addition to the Milwaukee Art Museum is one of his most powerful works to date, and a fitting homage to Eero Saarinen, who designed the original museum building on the shore of Lake Michigan. Unlike the Finn's famous air terminals, the Milwaukee museum is a low concrete monolith which by the 1990s was thought to lack architectural identity.

Calatrava's addition incorporates new galleries that are comparatively neutral, reserving its fire for the ceremonial spaces. The most striking feature is the giant brise-soleil, which opens and closes like the wings of a bird, serving as a sign to advertise exhibition openings. Writing in the *New Yorker*, the critic Paul Goldberger observed that this feature alone 'has made Milwaukee feel rather good about itself'.

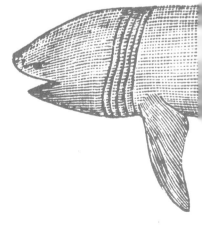

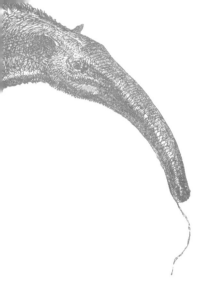

The Lyon Airport Station building is another example of Calatrava's flair in fulfilling a brief to provide a regional gateway. The ostensible purpose of the station was to connect the airport some 20 kilometres outside Lyon with both local and high-speed train services. But its job was also to symbolize the economic prospects of the region opened up by the new transport hub.

Calatrava was able to realize his vision without compromise in what became a notoriously expensive project even by the standards of major public works in France. The basic composition is of two intersecting halls, the lower platform hall, a 500-metre concrete-arched avenue with access to the trains, and, straddling this at right angles, a spacious glazed hall of triangular plan spanning some 120 metres. This space is clear, but leads to a concourse containing ticketing facilities, shops, transport offices and a restaurant, and gives pedestrian access to and from the airport.

The formal references to Saarinen are even more explicit than in Milwaukee. The metaphor of flight contained in the projecting curved parapet of the upper hall is expected; less so is its snouting profile from the side, which calls to mind a porcupine or anteater, an effect accentuated by the building's black and white colouration. From other angles, the building assumes a more insectile appearance, while inside some of the detailing of the glazing recalls the tracery of dragonfly wings. The structural form, so rich in biomorphic allusions, originated not from precursors in nature but from Calatrava's sculptural studies based more on the exploration of the physical balance of masses and forces.

Lyon Airport Station
Satolas, Lyon, France
1989–1994

left: A bird provides the most obvious and apt animal metaphor in Calatrava's Lyon Airport Station, but it is just one of many in a richly allusive building. From the side, the burrowing proboscis and the striped body of the glazed hall inexplicably conjures up the image of an anteater

right: Other animal metaphors compete with technological and architectural motifs ranging from the cockpit of a fighter plane to the complex vaulted roofs of medieval cathedrals. The concrete and steel structure of the upper hall (top) produces a rich interplay of geometric and plantlike shapes, while the concrete arches of the lower concourse (bottom) are more specifically like an animal's ribcage

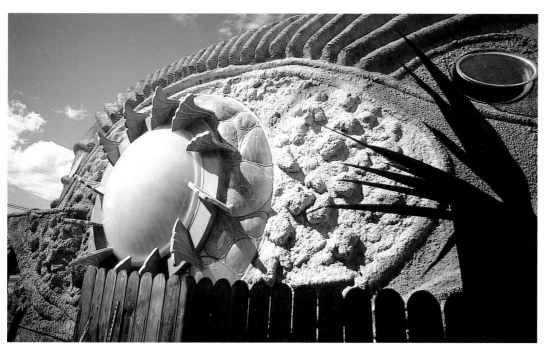

left: The concrete
finish gives a
unified appearance
to a house that is
loaded with the
suggestion of a
bizarre range
of animals

below left: The
organic shaping is
equally thorough-
going inside the
house, where
swooping and
spiralling curves
spin off from a
number of large
circular windows

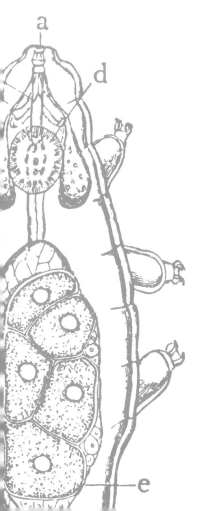

Eugene Tsui is an American, and perhaps especially a Californian, maverick. Paradoxically, this does not leave him out on his own, but locates him securely within a thriving tradition. He continues the vernacular explorations of Bruce Goff, who created dwellings with elaborate forms, both geometric and organic, using cheap, off-the-peg materials. To this, Tsui adds his own specifically naturalistic vocabulary as a means of expressing his commitment to 'evolutionary architecture'. His buildings contain elements of Fuller's Dymaxion designs, Expressionism and Art Nouveau. Zoological references abound – shell forms, bug eyes, scales and fins.

The house Tsui designed for his parents is loosely based on the morphology of the tardigrade. Tardigrades are a minor phylum of microscopic invertebrates also named water bears by the 19th-century English naturalist Thomas Huxley. They have four pairs of stumpy legs but are most closely related to certain worms. The creatures can survive extremes of drought and temperature – even down to near absolute zero – to live for up to a century. (They can even resist X-rays, and scientists have speculated that this capability, once understood, could be adapted for human travel in space.) Tsui draws attention to the geometry of the tardigrade's carapace – elliptical in plan with upper parabolic and lower catenary curves in section – which may be a factor in its resilience.

The house is also elliptical in plan. The lower structure of concrete foundation, cement-block and poured-concrete walls is partially buried. The roof comprises a series of sprayed concrete parabolic arches on a stressed wood frame. Like the tardigrade, it is designed to withstand extremes of temperature and shocks such as earthquakes.

The Reyes House was Tsui's first realized example of evolutionary architecture, which he defines in his book of that title as 'an architecture that implements the evolutionary practices of nature as a synthesis of billions of years of evolution applied to immediate needs and circumstances of form, function and purpose'. The project involved alterations to the interior of an existing house and the addition of a new space, which is roofed with a pair of six-metre-long translucent fibreglass wings like those of a dragonfly. These may be opened to dramatic effect for ventilation.

Reyes House
Oakland,
California, USA
1991–1993

Tsui's examination of the wing action of a dragonfly specimen during studies for the design determined structural and functional details such as the orientation of the major and minor struts reinforcing the wing panels

Watsu Center International School
Middletown, California, USA
1991–2001

The Watsu Center, a school for hydromassage, comprises five spherical modules like the structures made by honeypot ants clustered round two outdoor pools. The spheres are based on wooden frames and are sheathed in cement and waterproof vinyl composite. The shape has the minimum surface area for the volumes enclosed, which makes it thermally as well as structurally efficient

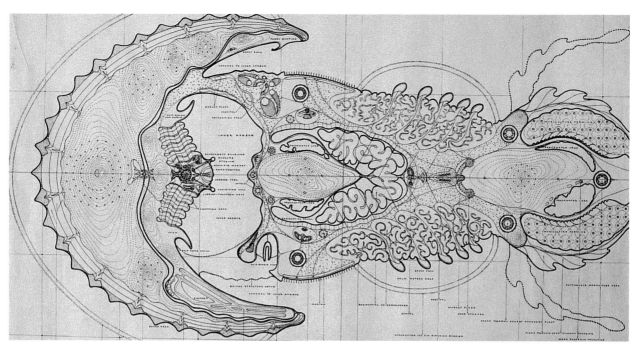

left
**Nexus Mobile
Floating Sea City**
1986 (concept)

The Nexus floating
city has a protec-
tive head with
mountains to
deflect tidal waves
while the limbs are
used to grow food
for a population of
up to 100,000.
Such requirements
influenced the
shape of the plan
to the point where
it began to suggest
a trilobite, the
outline of which
then became the
central motif

Tsui's hypothetical projects show no diminution of
his explicit animal symbolism. The Nexus sea city, a
proposed community of 100,000 people that recalls
Jules Verne's Ile à Hélice, is designed to be both
self-sustaining and self-governing, floating in waters
beyond the jurisdiction of national governments
outside the international 12-mile coastal limit. With
its fusion of political and formal idealism, it falls
squarely within the tradition of utopian designs from
Sforzinda, the 15th-century Italian model for many
utopian writings, and later Palmanova, to
Buckminster Fuller's air-floating cities.

Tsui, however, discards the circular or spherical template of these
utopias in favour of an animal plan
– a trilobite – that connects with
more primitive traditions.

The Nexus city's armature
would be constructed of concrete
made from dissolved salts in sea
water and deposited electro-
lytically on a sunken steel frame
in a process somewhat analogous
to the way many marine animals grow their skeletons. When sufficiently encrusted, the structure
would be floated to the surface for the completion
of the surface architecture. The 'head' of the trilobite supports a wave-deflecting mountain range,
while hydrodynamically extensible limbs provide
sufficient agricultural land to support the city's
population as it grows.

The city's mobility means that it would not
exhaust the ocean's resources in any one place but
move on in nomadic fashion, allowing the area left
behind to recover. Energy is generated from a
variety of sources including solar cells and wind
turbines, and these power desalination plants that
provide fresh water. 'The city is in essence a living
organism based upon ocean resources and climate,'
says Tsui.

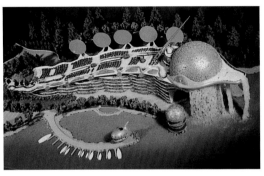

**Sky Dragon
Apartment
Complex**
Shenzhen, China
1999– (in design)

The Sky Dragon
complex contains a
range of split-level
two- to five-
bedroom apart-
ments and is
loosely organized
on the plan of a
garfish. The apart-
ments are reached
by a series of
ramps which lead
to a concrete shell
dome at the head
of the building,
which is used
for communal
purposes

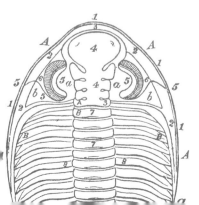

Michael Sorkin Studio

Animal Houses
1989–1991
(concept)

Michael Sorkin's
Animal House
series of projects
shows how readily
animalism may
be brought out.
In early study
models, a high
window and pilotis
are sufficient to
suggest a head
and legs, while
other elements of
the compositions
suggest the bulk
or motion of an
animal. With
subsequent devel-
opment, the
animal similes –
here to dog and
frog – begin to be
submerged under
a tide of more
conventional archi-
tectural readings

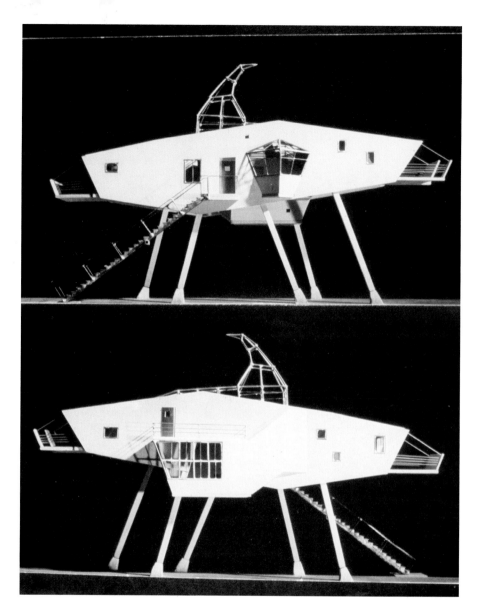

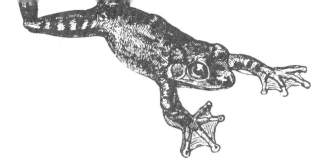

Michael Sorkin, a New York architect best known for his writings as a critic in the *Village Voice*, is notably uncomplicated about his habit of basing his own projects on animal forms. He does not offer elaborate functional justifications or hint at some obscure personal symbolism. He notes the near-universal 'anxiety that all representation is kitsch', but will have none of it for himself. And so he allows himself the pleasure of imagining buildings that are brazenly biomorphic. For, as he says, imitation is our primary source for form, and at root we can only imitate nature.

Sorkin sees past the rhetoric of styles based on moments in architectural history (neo- anything), or celebrating technological achievement in other fields (building as ship or machine), or derived from ideas in physics and philosophy (Deconstructivism). 'Such appeals to authority accomplish nothing if you're not already in thrall,' he suggests, 'which is not to say that the forms can't be fascinating and serviceable, just that they're no more inherently meaningful than the snout of the family dog.'

And the snout of the family dog is pretty much what has detained the Sorkin Studio for the last decade or more. The main polemics are two series of houses – hypothetical 'Animal Houses', individually named Dog, Frog, Aardvark and Sheep, and the more amphibious 'Beached Houses' named Carp, Slug and Ray, commissioned by a Manhattan art dealer as the beginnings of an artists' colony in Jamaica. The latter are studies in bilateralism, examining architecture's habit of symmetry and what happens as it is broken down. Animals such as fish 'are ideal research subjects because they are symmetrical but only until they wiggle. Our effort is to measure the space between the fish and the wiggle,' according to Sorkin. 'This is the study of a lifetime.'

Beached Houses
White House,
Jamaica
1989–1991
(unbuilt)

The arrangement of the Beached Houses allows for a main room on each of two storeys, which provides access to spaces in the creature's extremities, such as, in the case of Carp (top left), bathrooms in its projecting fins. Slug (bottom left) is a ground-hugging version with a long clerestory window on the upper floor, while Ray (above) is surrounded by an undulating verandah

Godzilla
Tokyo, Japan
1990–1992
(concept)

This is an attempt to capture and reshape the disorder of Tokyo as perceived or imagined from the distance of New York. Sorkin had not visited the city at the time of the project, but nevertheless wished to extrapolate from his vision of the place. The result was a chaotic vertical eruption that demanded the name Godzilla

More recent projects continue to be informed by these experiments in biomorphism. Increasingly, however, Sorkin's work finds itself connecting with practical requirements in unexpected ways as biomorphic and programme-driven agendas are found to overlap in terms of form and process.

In a project for a theatre, for example, Sorkin comes dangerously close to generating animal form out of the functional programme rather than his usual whimsicality. The prime requirement was that the theatre, for a company of puppeteers, should be transportable. The preferred form of the theatre – a roundish space uninterrupted by structural supports – led as it does in many such spaces to a certain likeness to a tortoise or turtle shell. In this case, a circular fabric membrane roof provides the shell which is supported on a demountable frame.

A series of amorphous floating islands, proposed as public art for a waterway in the centre of Hamburg, similarly finds its biomorphic formalism suddenly confronted with a legitimate rationale. A fleet of 'life boats' set to drift on wind and tide are as irregularly shaped as algae, but are also intended to perform some of these organisms' biochemical functions, since they contain machinery and chemical processing facilities to improve the water's cleanliness. The House of the Future extends this concept for dwelling units in a project designed to reflect the demographic shift away from ideal nuclear family life in America. The commune of double units includes facilities for 'bio-remediation' in an effort to ensure environmental sustainability.

Sheep
New York City,
New York, USA
1991 (concept)

Sheep – a good enough label for a squarish building on stumpy legs – was devised for a loft development on a vacant lot in SoHo. Sorkin's plan argues against the alluring but useless 'flexible' open plan that the 'loft' represents by providing 'particularity' in a wide range of rooms and spaces of different sizes and configurations

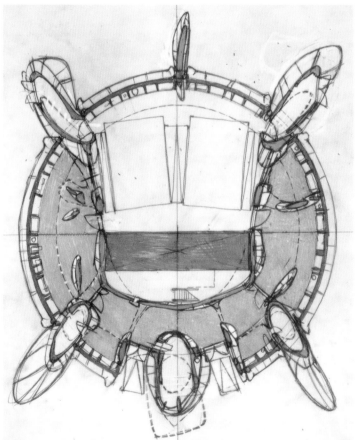

House of the Future
1999 (concept)

Sorkin's proposal to replace the typical American family house with something more in line with demographic trends takes the form of double living units ganged together to form a colony or 'house'. Each 'house' shares communal working and dining space and recreational and children's spaces in a manner reminiscent of the social idealism of early Modernism

Turtle Portable Puppet Theater
1995

Sorkin's interest in animal architecture dovetails nicely with the programme of this demountable theatre building. The functional demand for unobstructed space is met by the fabric-covered 'turtle' shell, an animal resemblance accentuated by articulating the exits as the creature's legs

Morecambe is a seaside town of somewhat faded charm a few miles up the coast from Blackpool, Europe's most visited leisure attraction. Needing to strengthen the seafront against coastal erosion, the town council held an open competition inviting proposals for revitalizing the town based around these vital features.

Birds Portchmouth Russum's brash response far exceeded the brief, and in effect demanded to know how much revitalization the town councillors really wanted. Reasoning that the breakwaters were necessarily obtrusive, the architects chose to make them a monumental and joyous feature of the coastline. Inspired by the bright lights and massive steel fairground structures of Blackpool, and more generally by high Victorian seaside extravagance, they proposed a series of four brightly coloured giant shrimps – one for each of the villages that historically amalgamated to become Morecambe, and each to be built on existing breakwater foundations. The audacious series of structures – a marina, an amusement arcade, a theatre and concert hall, and a lifeboat station – would be strung out like glittering gems along a tree-lined boardwalk dotted with refreshment kiosks and entertainment stalls.

The structures are quite obviously shrimps – a homage to the local delicacy. But they also read as architecture: Constructivism, Archigram and even James Stirling are not far away. However, this astonishing *jeu d'esprit* was never built, and Morecambe has continued to fade.

Birds Portchmouth Russum

Seafront Redevelopment
Morecambe, England
1991 (competition scheme)

The massive scale of the engineering, the polychrome finishes, and the cavalier approach to matters of taste all smack of the heroic period of high Victorian architecture when the 'seaside' was popularized. This imaginary construction drawing echoes Brunel's drawings of his great engineering feats. Four such structures house very different functions but are united by their resemblance to Morecambe Bay's famous shrimps (see pp. 38–39)

Plashet School Bridge
London, England
1999–2000

The essential structure of the bridge is provided by a giant curving carriage beam made up of vertical I-beams which form the balustrades and a steel deck. The blue-painted steel-plate bridge supports are cut into the shape of cupped palms, providing a human gesture of strength and care. The asymmetric steel supports for the fabric covering provide interior spaces (right) of different character at points along the bridge according to whether they are oriented in the same direction or not

Birds Portchmouth Russum's animalism is less specific in this completed scheme for a covered footbridge linking school buildings on either side of a busy road. The snaking form derives from the constraints of the project, which demanded an S-bend in order to achieve axial alignment with each building and to steer around a mature tree in the school grounds.

Although simple 'covered wagon' hoops would have provided a functional covering for the 67-metre bridge, the architects chose asymmetric steel arches to add visual interest. On straight sections, these alternate, giving the bridge a swaggering appearance, as if elbowing its way forward. This arrangement also maintains the double curvature needed to hold the Teflon fabric steady in tension. On the bends, the arches are oriented in the same direction, giving a more spacious feel near the midpoint of the tube, where there is a vantage point with seating.

In visual terms, the concave segments are like the vertebrae of a curving spine rather than the segments of a worm's body. The resemblance to a worm is closer structurally, although with an important difference. Each segment of a worm's body has a layer of longitudinal muscle and a perpendicular layer of circumferential muscle which enable movement. The layers are maintained in slight tension between each partition by the pressure of body fluids in the worm making each segment bulge out; in the bridge, however, the situation is nearly reversed, with no pressure differential, and the membrane is pulled in from the hoop frames that support it.

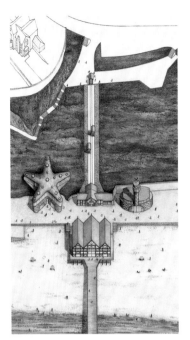

Seaside Café and Lifeguard Station
Saltburn, England 1998 (competition scheme)

The rare opportunity for the public to see a building from above gave the idea to base the plan of this café at the foot of a cliff on a starfish. The café is situated to one side of a cliffside tramway leading out onto a pier, and is balanced by the lifeguard station, which is elliptical in plan. The fantasy aspect of the buildings is accentuated by their brass and copper finish

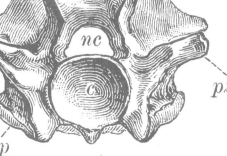

Two aboriginal centres in remote locations in Australia are a powerful reminder of the role originally played by animal representation in human culture. The Brambuk Living Cultural Centre in Victoria takes its plan outline from the cockatoo with wings outspread, that bird being the totemic symbol of some of the Koori tribal communities the building serves. The Uluru-Kata Tjuta Cultural Centre near Ayers Rock (the aborigine name for which is Uluru) draws its symbolism from snakes important in the mythology of the Anangu people.

The architect of both buildings is Gregory Burgess, who won the jobs because of his practice of arriving at architectural solutions through community workshops. The stories of the indigenous communities enrich the architecture, but are not translated in the most obvious fashion. At Uluru, for example, Burgess noticed the hand gestures of an Anangu woman telling the story of the snakes Kuniya and Liru. These gestures were explored in the workshop by drawing in the sand, leading to a leitmotif for the centre. In addition, contemporary Western narratives to do with environmental sustainability, local materials and vernacular tradition are woven through the buildings.

In the aboriginal stories, the animals have dual existence as persons or tribes. This property of metamorphosis, traditionally communicated through dance and other ritual, is central to the stories' power and indicates the intensity of the indigenous communities' connection with the animal world. The potency of the architectural symbolism depends not on transparent obviousness, but on multiplicity and ambiguity. Just as a single species may act as the totem for several tribes, or metamorphose between animal and human guises, so Burgess's aborigine centres incorporate overlapping animal messages – more in fact than their architect intended, for some meanings were encoded when the aborigine communities inaugurated the buildings and conferred totemic significance on aspects of their design.

Uluru-Kata Tjuta Cultural Centre
Uluru-Kata Tjuta National Park, Northern Territory, Australia
1990–1995

Ambiguous symbolism can have a universal potency. At the Uluru centre, the snakes central to aborigine mythology are highly abstracted in the design, and mixed with other motifs. They retain their authenticity for the Anangu, but the language also communicates to visiting tourists by leaving room for people's imaginations to roam

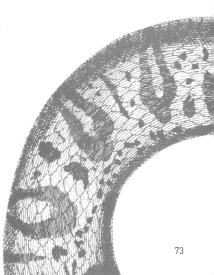

The cockatoo, loosely indicated in the feathery undulation of the corrugated iron roof, is just part of the story at the Brambuk centre. There are also readings of the emperor moth and the eagle Bunjil, the creator god in the local aborigine theology. Bunjil presides particularly over ceremonies governing tribal intermarriage, and as such could be said to bless the mixing of the indigenous Koori tribes and the tribes of Western tourists at the Brambuk centre. Different tribes found different symbols in the completed building: a forest tribe identified with the tree trunks used as columns, a coastal tribe found whale forms in the structure, while another tribe interpreted the spiral ramp to the upper floor as an eel, a traditional staple.

Burgess's work for the Koori communities helped secure the project at Uluru. According to the local mythology, in the time of the creation, Kuniya, the non-poisonous carpet python, migrated to Uluru where it lost a great battle to the venomous Liru. Following the creation period, the Kuniya snake-people were turned to stone. The rocks that bestrew the landscape today represent persons or artefacts of the Kuniya camp, while the red striations of the stone denote the blood of that battle. The plan of the Uluru centre represents the two snakes eyeing each other across a ceremonial court-yard with views to Ayers Rock. As at Brambuk, the ground-hugging two-storey building responds to the local geology and contains exhibition space and other visitor facilities.

Even without the excuse of aborigine mythology, a number of Burgess's buildings display animal motifs, reflecting the architect's belief that such symbolism has the power to bring about what he calls a 're-enchantment with place' and to signify connection – or, in the case of Western societies perhaps, reconnection – of humankind with the natural world.

top right
Brambuk Living Cultural Centre
Grampians National Park, Victoria, Australia 1986–1990

Located beside a creek in a spectac-ular valley between the two ranges of the Grampian-Gariwerd moun-tains in western Victoria, the Brambuk centre's ground-hugging profile points up the dramatic sur-rounding landscape

below right
Twelve Apostles Visitor Centre
Port Campbell National Park, Victoria, Australia 1998–2001

Burgess responds with images of whale, ship, bird and leaf to the real and imagined force of the elements at this exposed site

far right
Heide Rose Pavilion
Bulleen, Victoria, Australia 1991

The bird's-wing motif reappears in a suburban garden pavilion

The first phase of the Darwin Centre at the Natural History Museum in London already houses 22 million specimens of vertebrate species gathered by Charles Darwin on his voyage aboard HMS *Beagle* in the 1830s and by other expeditions before and since. The second phase will provide similar display and storage space for 28 million insects and six million plant specimens.

Appropriately for an entomological archive, the design resembles a giant cocoon slotted into its own specimen box. But this was not how the project began. The irregular ellipsoid at the core of the building developed not with any zoomorphic idea in mind, but from a functionalist rationale. The site is squeezed between existing museum buildings and is taller than it is broad. The first thought was that a large atrium should form the focus, but this was dropped because it was wasteful of display space. Instead, it was decided to insert a sculptural form into the rectangular void. The project architect, Anna Maria Indrio, opted for a single large shape of complex curvature for reasons to do with perceptions of scale and proportion. This form contains eight floors of galleries, making it possible to put on view the bulk of the specimen collection. Museum visitors gain access to the displays from bridges on one long side of the cocoon, and may emerge across bridges on the other side to encounter the museum's scientists at work.

It was found that the intended smooth plaster finish of the cocoon could not be achieved without risk of cracking. But a revised design made a virtue of this setback by cutting the cocoon into sections and then fastening them together with steel belts wound round in a way that looks natural rather than geometric. This structural embellishment emphasizes the form's resemblance to a cocoon woven from silken thread.

C.F. Møller and Partners

top far right and bottom: The cocoon-like central element of the Darwin Centre is just visible through the curtain-wall exterior of the new building, which symbolizes protectiveness as well as hinting at what it contains. Visitors within the museum on the other hand are too close to the form to perceive its totality; instead, say the architects, 'the mind completes its geometry'

top right: The surface of the sprayed concrete cocoon is wound round with steel reinforcement. This structural – and decorative – detail appears random, as if spun by a silkworm, but is in fact calculated to distribute the stress evenly across the complex curved surface

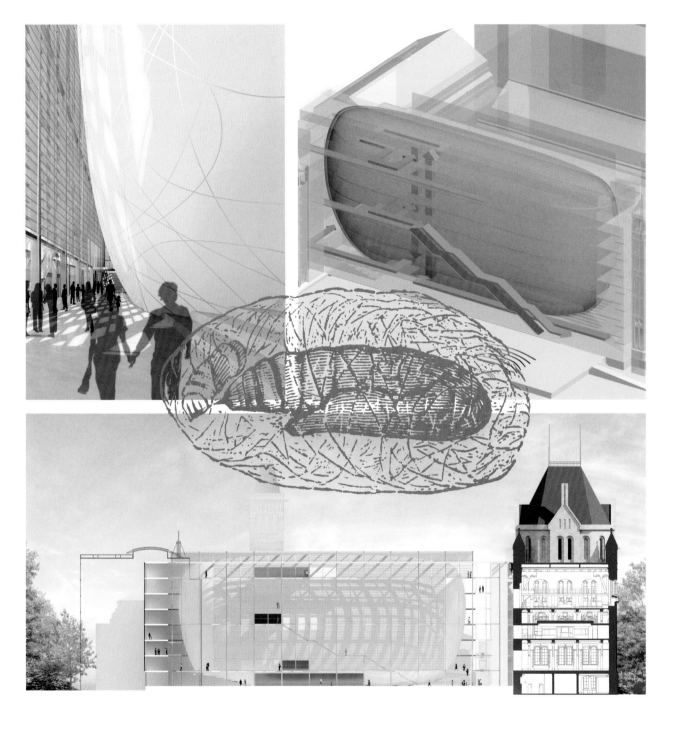

Snøhetta and Stephen Spence

Turner Centre

Margate, England
2001–2005

The deliberately ambiguous shape of the timber-clad concrete Turner Centre gallery building – is it a whale, a shark's fin, a submarine conning tower, a pebble? – is intended as a magnet to draw the curious to the town and catalyse the area's economic regeneration

Joseph Mallord William Turner famously lashed himself to the mast of a ship in order to capture more accurately in oils the sensation of a storm at sea. The artist's experimental approach and disorienting canvases became guides for Snøhetta and Stephen Spence as they approached the commission to design a new centre dedicated to the artist in the town where he spent much time.

Turner returned to Margate throughout his life and painted many seascapes there. These paintings are among his most benign works, and the innovation that he saw all around him in the world, which appears as both exciting and disturbing in his masterworks, is gently embraced into the Margate scenes showing steamers in calm seas. Snøhetta must hope that the abrupt modernity of the Turner Centre will slip as happily into its context.

The centre is in two contrasting parts linked by a covered glass bridge – a low-slung pavilion on the stone pier, housing a ticketing foyer, restaurant and shop; and, rising out of the water to seaward of the pier, a 30-metre-high curvaceous but enigmatic concrete mass containing the display galleries. The elusive 'pebble' shape and intended grey weathered-oak finish of the latter will play optical tricks. Sometimes a bulk silhouetted against the sea like a whale coming up for air or a giant shark's fin, the building in other lights will dissolve against the sky like the confused horizons in a Turner painting.

Three galleries will exhibit the work of contemporary artists as well as the paintings of Turner and his peers. Two of these galleries are conventional, while the upper, main gallery is designed as a fluid space without clear delineation between walls, ceiling and floor. Ramps up to the main entrance and within the gallery building prepare visitors by creating a sensation of movement and disorientation that was so much part of Turner's vision.

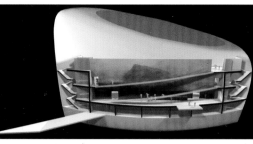

Ramps provide the approach to the Turner Centre as a whole and to the galleries within, making for ease of navigation but also ensuring that visitors lose their Cartesian reference points before they view the paintings. The principal gallery is a large ovoid illuminated by natural light from above. Sharp corners where walls meet floors and ceilings are rounded and the paintings will be displayed pulled forward from the interior surface so that they appear to hover in space

animal by function – statics

Zoom in, like Charles and Ray Eames's famous film *Powers of Ten*, on one of Buckminster Fuller's geodesic domes. Hold the concept of the shape in the viewfinder, but increase the magnification, and one will find objects in nature with the same polyhedral structure and symmetry time and again as one goes down to smaller scales. The marine protozoans known as radiolarians have it, so do some viruses, so, even, does the carbon molecule buckminsterfullerene. Try the exercise again with a covered sports stadium, and one sees the shells of turtles, molluscs and insects. These are basic forms, to be sure, but they illustrate a general principle famously elucidated by D'Arcy Thompson that some of the best structures achieved by man are also seen in nature for the simple reason that both building and organism must respond to the same physical forces.

During the last century, the architects who have considered themselves the champions of structural integrity paradoxically appear almost to have ignored the living world. Their work has been not organic but defiantly inorganic, based on repeated identical units arranged in potentially infinite lattices. This is the principle behind buildings from the Eames House to Norman Foster's Stansted Airport.

Shed and blob might seem poles apart at first, but there is in fact a progression from one to the other that may be explained with the help of intermediate examples. The simplest shed is based on a flat grid, its structure extended by modular units along two orthogonal axes to form a cuboid. The first step towards apparently organic form is to bend one of these axes into a circle. An open-ended arched structure is the result, a longitudinal slice of a cylinder. Now, bend the second axis. If it is bent to the same degree, a spherical dome is the result; if not, the form is elliptical rather than circular in plan, and is a segment of a toroid. With each step, some of the economy achieved by repeating parts is sacrificed, but not as much as one might think, since the underlying mathematics is still simple. This level of complication seems to represent the current state of the art, and a surprising number of new buildings today are based on toroidal segments. But it is possible to introduce further distortions to the geometry to arrive at 'organic' morphologies that effectively disguise their mathematical roots, just as nature has done for example in the shells of snails and molluscs, which manifest a rich variety of conical, spiral and flattened forms, but are all derived from no more than three growth vectors.

Even the illusion of natural development seems to strike a chord in people. If, as Christopher Alexander argues, there are urban environments that appeal because of the sense they give of being organic, then surely the same may be said of individual buildings. A city or a building that appears to convey a sense of its own ontogeny acquires a feeling of inevitability, and perhaps even a 'rightness'. Such a building may appeal to people even if they cannot immediately locate a natural analogy or identify the structural principles it shares with living organisms.

Critical wisdom pitches the simplifying Modernist against the complicating organicist. But these developments expose this dichotomy as a dead letter. The feud is resolved at a stroke by an architecture that extends the possibilities of 'high-tech' with one hand, and with the other gives the organic school its first chance of large-scale realization.

Renzo Piano
Building Workshop
**Padre Pio
Pilgrimage Church**
San Giovanni
Rotonda,
Foggia, Italy
1991–

Wilkinson Eyre Architects

Multiplex Cinema
Merry Hill, Dudley,
England
1998– (on hold)

Twenty auditoria
are stacked two or
three high, these
stacks forming
eight segments of
decreasing size
within the overall
spiral plan. The
Fibonacci sequence
found in natural
spirals and the
scrolls of Ionic
capitals provides
inspiration here,
although its num-
bers mount up too
rapidly for them to
correspond exactly
with the sequential
theatre seating
capacities

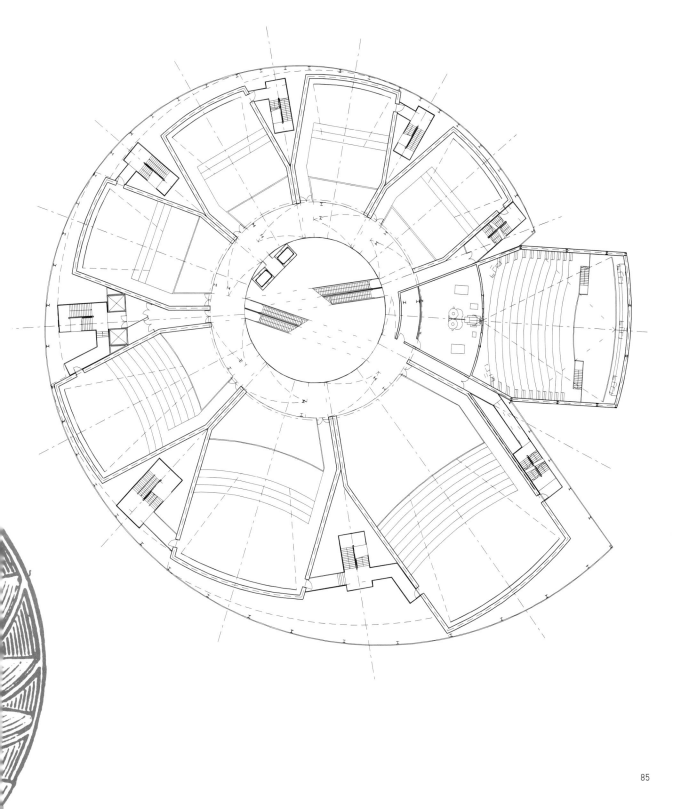

Multiplex Cinema

The spiral design is apparent in plan, but is not an important visual feature for cinema-goers, although it does make it easy for them to navigate through the building. The visual and functional resemblance to nature breaks down at various points. The nautilus has no central void accessing the chambers that it uses for buoyancy, for example, and where the nautilus has a large open cavity housing the animal itself in the outermost chamber, the cinema building has a video wall, with entry made through a point in the spiral shell wall

Wilkinson Eyre Architects' plan for a multiplex cinema in a suburb of Birmingham shows clearly how the programmatic requirements of a building can lead to a formal solution that shares certain qualities with natural organisms. In order to accommodate 20 theatres, the architects chose to distribute them in sequence by seating capacity around a central atrium, which provides a common lobby space and access to each of the theatres off a spiral ramp.

The spiral has a mathematical basis in the Fibonacci sequence (a series of numbers in which each term is the sum of the two previous terms: 1, 1, 2, 3, 5, 8... and the ratio of adjacent terms approaches the Golden Section). The sequence is found in nature in flowers and seed cones and in the proportions of the sequential chambers of the shells of some snails and molluscs such as the giant nautilus. The spiral is a rare occurrence in architecture, with a couple of notable exceptions from the 1950s in Bruce Goff's Bavinger House and Frank Lloyd Wright's Guggenheim Museum.

The spiral solution here provides cinema-goers with an easily navigated environment, unlike that of a large old cinema that has been divided up, or even a conventional multiplex housed in a standard shed. The radiating auditoria are wrapped in a smooth outer wall, with the cladding set at an angle to hint at the spiral plan that is otherwise hard to see in the elevation view. The individual cinema 'chambers' may be discerned through translucent strips in the wall in an echo of the delicacy of the nautilus shell. 'Even though viewers might not see the natural analogy or the structural principle,' Jim Eyre suggests, 'they can appreciate that it looks as if it has grown, and can connect deeply with that.'

The repetition of structural parts, not in a straight line as in engineering sheds from Brunel to Chris Wilkinson's own early work, but here radiating from a centre point extends the possibilities of building form within the orthodoxy of 'high tech'. The solution in the case of this retail unit creates a local landmark, while remaining potentially competitive with standard design solutions.

Retail Warehouse
Merry Hill, Dudley,
England
1998–2000
(unbuilt)

The radial structure of the shed and the column-free space within it carry the sea urchin analogy beyond the merely visual

The architects experimented first with a pebble shape modelled in clay, the geometry being regularized as engineering and cost considerations were taken into account. The resemblance of what was now a squashed sphere to a sea urchin was immediately obvious. Structurally, too, the building is not unlike these echinoderms, whose bodies consist of a central cavity protected by an external skeleton of fitted plates of calcium carbonate. Here, clear interior space is provided beneath a roof of laminated timber ribs radiating at 15-degree intervals covered with plywood panelling. The wooden frame is covered with a waterproof membrane, and this is in turn covered by a stainless-steel mesh over glass spacers. Although rationalized on functional grounds because three-dimensionally curved cladding was prohibitively expensive, this textured surface serves to reinforce the animal image of the building.

A painter as well as an architect, Wilkinson is aware of the danger of allowing the functional programme to dominate the design to the exclusion of all else, and is determined to allow intuition to play a part. Perhaps this is why the sea urchin idea is abruptly curtailed to provide an inclined, curved-glass showroom façade that slices through the spheroidal envelope of the building, compromising both the visual and structural analogy with nature.

The sea urchin form of Wilkinson Eyre's retail warehouse dominates from some angles, providing an unexpected intrusion amid dull surroundings of retail sheds and car parks. From the front, however, it is apparent that the architects have subverted the analogy with nature by cutting into the shape with a glazed wall. For Wilkinson this act is an important assertion of the primacy of the architect

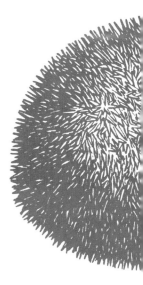

Car manufacturers, and German car manufacturers most of all, are among the most enthusiastic patrons of biomorphic architecture. Their adoption of the style is one of the surest indications that it is set to become a major feature of the cultural landscape. Although it may not be explicitly stated as a requirement in the briefs that these clients give, their architects are almost universally interpreting their needs by using this new vocabulary. It is impossible to disentangle whether this is an architectural response to the evolving shape of cars, or whether it is due to the uptake of computer technologies that facilitate the development of fluid forms (and which themselves have influenced car shapes), or whether it is a reaction to a more general cultural mood of environmental awareness that, in the case of Audi, BMW and Mercedes-Benz, is undoubtedly borne along on a strong undercurrent of German Romanticism.

In their presentation to Audi UK, Wilkinson Eyre showed both a manta ray and a stealth bomber in an effort to stimulate interest in the asymmetric aerofoil flat curve, a form that they feel has been neglected in architecture, despite its potential to reduce internal volumes and introduce visual subtlety. But it was Audi's TT sports car rather than these examples that first set them off on this train of thought. Although historically difficult to generate because they are not made up of circular sections, these curves can now be created and fabricated more cost-effectively with the help of computers.

Audi UK Regional Headquarters
London and Solihull, England
2000– (competition scheme)

The subtly curled lip at the midpoint of the single-curvature sloping aluminium roof, together with its raked profile in plan, suggests the natural hydrofoil of a manta ray as well as aeronautical motifs

Magna Science Adventure Centre
Rotherham, England
1998–2001

Magna is a science centre based in a vast redundant steelworks. Funded by the UK Millennium Commission, it takes as its themes the Aristotelian elements, earth, air, fire and water. Wilkinson Eyre responded by inserting distinctive pavilions for each element. The organic forms provide an effective counterpoint to the severely rectilinear hall (see also pp.80-81)

*Renzo Piano Building
Workshop*

**Auditorium Parco
della Musica**
Rome, Italy
1994–2002

Supported on brick
plinths, the lead-
clad forms of
Piano's concert
halls appear to
float above the
ground, an effect
that enhances
their metaphoric
power, whether
the metaphor be
musical or animal.
The expressed
hard carapace
implies the
presence within
of a space
uninterrupted by
structural supports

The architecture of Renzo Piano achieves a synthesis of themes that concern many architects individually but that are seldom brought together with such aplomb. The technocratic, and indeed technophile, streak of his breakthrough building, the Centre Pompidou, still runs strong in his work, notably in the recent Kansai Air Terminal. An ability to bring vernacular materials into this idiom in turn leads to architecture that displays a rare sensitivity to regional context; add to this a more convincing environmental agenda than that of many architects. These ingredients are mixed to produce buildings of cool sophistication that have no need to clamour for attention.

Piano is adept at softening the unremitting machine aesthetic of 'high tech' by his attention to materials and details. And it is here that a perhaps surprising naturalism emerges, not so much by design but as the inevitable consequence of the chosen architectural solution's being the sort of solution that nature might arrive at in similar circumstances.

The Menil Collection Museum in Houston, for example, one of Piano's most satisfying buildings, is a highly serviced shed in the 'high-tech' tradition. A system of ceiling louvres diffuses the harsh Texas sunlight to create a placid gallery environment. These louvres are slung from triangular space-frame trusses. As D'Arcy Thompson shows, the metacarpal bones in the wings of some large birds such as certain vultures incorporate similar space frames for greater lightness (or, more precisely, for a greater power-to-mass ratio, which is not a problem for small birds). In the hands of another architect, these features might have remained a brutal mechanism, and the natural likeness would have been obscured. But, as with the tapering steel brackets or 'gerberettes' that support the external skeleton of the Centre Pompidou, Piano's castings are so shaped that the resemblance to bone can hardly be ignored.

The Menil Collection Museum
(Piano & Fitzgerald Architects)
Houston, Texas, USA
1982–1987

Triangular cast-iron space frames form the roof structure of the single-storey galleries at the Menil Collection Museum. These trusses bear a close resemblance to the metacarpal bones in the wings of the pterosaur of the Mesozoic era and some other large flying species. The trusses support white-painted light baffles which are equally bony in appearance

Mercedes-Benz Design Centre
Sindelfingen, Stuttgart, Germany
1993–1998

Mercedes-Benz persuaded Piano to come up with a more curvaceous design after dismissing early plans as 'too Bauhaus'. The building provides further evidence that German car manufacturers feel the need to soften their image – and are doing it literally through their architecture. The somewhat mysterious form of the centre signifies the creative endeavour of the design team within, but also provides the necessary security for an industry paranoid about keeping its design secrets

Looser biomorphic images are exploited with clearer metaphorical intent in some of Piano's newer buildings. The Mercedes-Benz Design Centre is a freed-up version of a traditional factory, with the familiar sawtooth roofline softened and bent through an arc. Each roof plate is gently curved (it is a section of a toroid) and overlaps the next, with skylights interspersed between them. The fan shape suggests numerous parallels in nature from fishes' fins to coccoliths, microscopic chalk fossils, while Piano likens it to an open hand.

The Padre Pio Pilgrimage Church develops this roof concept of overlapping plates. Here the requirement was to provide a large, welcoming space for pilgrims – the church can hold a congregation of 6,000 people. In plan view, the structure is spiral, with some 20 unequal slim stone arches radiating from a central point like a spider's legs or the feeding tentacles of a fanworm. A single low dome of patinated copper panels is thrown over this complicated but expressive structure.

The Auditorium Parco della Musica is a complex of three discrete concert halls in parkland containing the buildings of the 1960 Olympic Games. Seating 2,800, 1,200 and 700 respectively, the halls nevertheless have a similar morphology dictated in part by acoustical requirements: each comprises seven exoskeletal lead-clad plates forming a long, flat tail section and, protruding from it, a smaller head section. Though 'conceived as veritable musical instruments', the buildings have strong animal suggestions about them, too. The elongated grey elevation is slug-like, while the armoured exterior plates suggest a scarab beetle, and the head protrudes from the body like that of a turtle. To Piano, the buildings seem like 'a metaphor of antiquity' – hardly needed as it turned out, since a Roman villa was uncovered during preparation of the site.

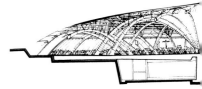

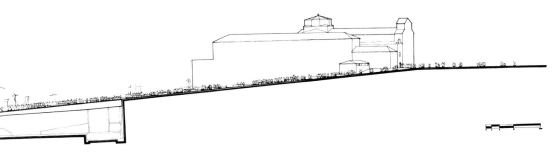

**Padre Pio
Pilgrimage Church**
San Giovanni
Rotondo,
Foggia, Italy
1991–

The low-vaulted
space of the Padre
Pio church pro-
vides a new focus
for visitors to the
monastery where
Padre Pio was a
monk. The stone
arches provide an
echo of the Gothic
brought up to date
by computer-aided
design. The radiat-
ing spiral of arches
supports a single
shallow dome
of patinated
copper panels

The 40-storey UK headquarters of the major reinsurance company Swiss Re is like no other in the City of London. It is claimed to be 'the capital's first environmentally progressive tall building', and it looks suitably different from its neighbours.

Foster and Partners

As one expects from a Foster project, there is a strictly functional rationale for the radical form. Most of the load of the 180-metre cigar-shaped building is carried by an external diagonal grid of steel beams stiffened by horizontal hoops. The diagonal matrix has allowed the architects to insert light wells that spiral up the building and break up each floor's circular plan into smaller areas. These wells permit air to circulate naturally, assisted by the aerodynamic envelope of the building, thereby reducing energy consumption. At ground level, the tapering profile leads to a reduced footprint and greater public space, while the circular plan streamlines wind flow.

The building's form is close to that of various sea sponges – and any hint of nature, of course, helps to advertise claims to environmental sustainability. Many small marine creatures that affix themselves to the seabed, among them sponges, anemones and *Foraminifera* (the word means hole-bearing or perforated), have calcareous or siliceous exoskeletons. As the animal grows, it builds up a tracery of sometimes amazing geometric regularity. These delicate frames are sufficient to support and protect the enclosed soft body of the organism in the gravity-neutral deep-sea environment.

The analogy extends to aspects of the building's function, too. Its shape ameliorates the wind flow, just as the sponge's shape helps water flow round it. Internally, the natural ventilation system finds a precedent in the way that sponges feed, sucking in water through basal membranes and expelling it through a chimney, or osculum, at the top.

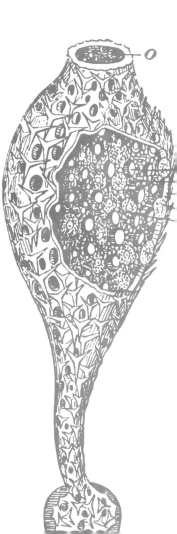

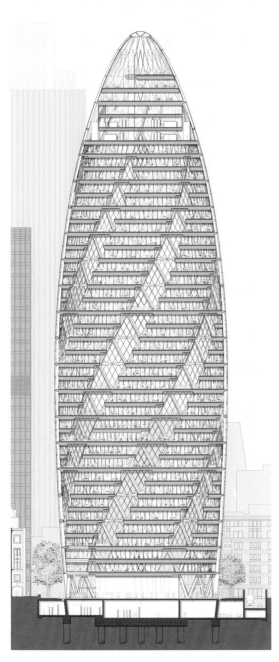

Swiss Re Headquarters
30 St Mary Axe
London, England
1997–2004

The elegant triangular-faceted structural skin of the Swiss Re headquarters has echoes of the domes of Buckminster Fuller and Bruno Taut, both of whom acknowledged natural archetypes for their forms. The tapering profile and circular plan provide aerodynamic conditions which reduce ground-level wind speeds compared with a rectangular tower, and assist natural ventilation internally by means of spiral light wells. Foster's building has visual and functional resemblances to species of sea sponge, but requires steel to carry off a structural echo of what nature does in their delicate mineral lattices

Buildings often acquire nicknames that play on their morphology. Fond or pejorative, they can become ineradicable labels for districts and even whole cities – think of the Flatiron building in New York or 'Pereira's prick' in San Francisco. For this to happen usually requires the building to possess a more or less obvious resemblance to some familiar object. Thus, Utzon's Sydney Opera House, for example, has prompted a catalogue of similes, many of them drawn from nature.

Foster's architecture, a million miles away from Utzon's expressionism, is perhaps the last place one would expect to find this name-calling. Yet even his buildings provoke this response, not because of their intended allusive qualities, but because their very lack of reference points from traditional architecture leaves people at a loss how to describe them. The

Swiss Re headquarters has been clumsily dubbed the 'erotic gherkin', while his early master-piece, the Willis Faber and Dumas offices in Ipswich, with its sheer serpentine black glass wall calls to mind a grand piano.

Foster's exhibition centre on the banks of the River Clyde in Glasgow was dubbed 'the armadillo' by a journalist, and the label has stuck. The name says little for journalistic powers of observation. For not only does it resemble this creature hardly at all (a giant prawn or a chiton would be closer), but so many other references are overlooked. The intentional reference is to the upturned hulls of the ships once made on Clydeside. The bands come to a Gothic point, an unexpected historical allusion. Other aspects of the building's detailing suggest machinery, aircraft hangars and industrial references more in line with Foster's other work. And if we are not meant to find an armadillo, then there is surely at least the suggestion of armour in the lapped, aluminium-clad roof panels.

Scottish Exhibition and Conference Centre
Glasgow, Scotland
1995–1997

Foster has discarded the old 'high-tech' ideal of the extensible shed for an organism-like wholeness and a distinctive profile on the skyline. The bizarre silhouette is sufficiently removed from our expectations of what buildings should look like to produce an uncontrollable flow of similes. The armadillo analogy, however, is compromised by the fact that the roof sections are lapped in both directions quite unlike that creature's armoured bands. Foster is said to dislike the armadillo nickname, but it is evidence that his building has succeeded in its objective to provide a civic symbol

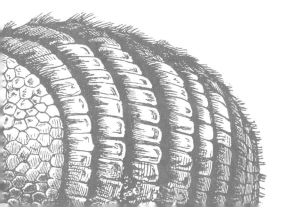

The Ecole des Beaux-Arts may have lost its fiat governing styles of architecture, but it continues to influence the attitude that architects take towards the other disciplines of the building art. Engineering structures such as the nets of Frei Otto or the bridges of Santiago Calatrava may be admired; but when these engineers start to do real buildings, architects start to cavil.

The American attempt to fuse industrial engineering and architecture led by the Eameses, Craig Ellwood and others after the Second World War soon withered, not least in America itself. English 'high tech' depended on creative and tolerant structural engineers such as Peter Rice, Frank Newby and Tony Hunt, who, some architects fear, may not be replaced by a new generation. But even here, the two disciplines maintained separate offices.

By contrast, Samyn and Partners are 'architects & engineers' under one roof. The Belgian company designs bridges as well as buildings, and Philippe Samyn says of himself that he is 'basically a structural engineer'. After the example of D'Arcy Thompson, he is engaged in his own theoretical research into the morphology of structures. He takes his lead from the engineering constraints that he perceives, and his purist solutions consequently often bear a pronounced resemblance to forms in nature.

The segmented tent-like structure that encloses the research laboratories of M&G, an Italian chemicals company, illustrates the process. The project had to house areas for both heavy and delicate research, and be adaptable to future shifts in the balance of these requirements. Samyn's solution was to erect a huge tent sheltering flexibly planned one- and two-storey laboratory 'buildings' and heavy plant. The membrane structure is not simply an extruded sleeve, but, stretched over a series of gently rising and falling space-frame hoops, tapers at each end to form a caterpillarish enclosure.

Samyn and Partners

**M&G Ricerche
Research
Laboratory**
Venafro, Italy
1989–1991

Samyn's buildings are entire engineering conceptions rather than architecture with a few bravura 'high-tech' flourishes using novel materials. Here, the purity of the form that results is accentuated by the bright white PVC-coated polyester membrane and its setting in a reflecting pool, and fortuitously happens to echo the arched greenhouses found in this part of Italy

In order to increase the height and usable enclosed area of the laboratories, the legs of the structural arches curl in and touch the ground in a way that evokes a crawling caterpillar, although the detail is driven by the engineering solution and not by any search for animal metaphor

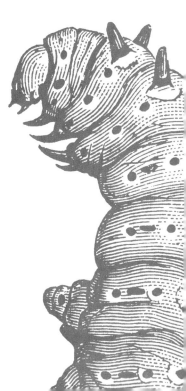

Others of Samyn's buildings arrive, like the M&G laboratories, at the blister shape of animal carapaces – elliptical in plan and circular-segmental in both sections – as the natural consequence of wishing to enclose the maximum area within a serviceable protective shell. The advantages of such a shape become apparent in a flat area of landscape such as that occupied by the Totalfina petrol station at Mannekensvere, one of a number of buildings that Samyn has designed for the Fina company. The low-rise form eliminates the need for deep foundations, so imposing less of an ecological burden on the site. It also reduces the wind loading, which is an important consideration in an exposed position. The building's appearance of a protective carapace is heightened by the choice of the cladding material – corrugated steel arranged in broad bands with expanded metal mesh beneath.

Totalfina Service Stations
Mannekensvere, Nieuport, Belgium 2000 (competition scheme)

Layers of corrugated and perforated metal sheeting used as cladding make comparisons with an animal carapace inevitable

Unquestionably the most beautiful of these purist objects in the landscape is the Walloon branch of Matériaux Forestiers de Reproduction, a futuristic structure of considerable mystery isolated in the midst of the Ardennes forest. The building houses offices, laboratories and cold-storage facilities used in seed preparation and woodland management. The success of the building relies on its marriage of 'high tech' construction and natural materials. The framework uses timbers of 200-year-old oak cut to a number of standard lengths and arranged in a two-dimensional grid to achieve economy of scale.

Comptoir Wallon des Matériaux Forestiers de Reproduction
Marche-en-Famenne, Belgium 1992–1995

Timbers cut to standard lengths allow the building to fulfil the ideals of industrialized construction more commonly achieved in steel and glass while also giving the appearance of belonging in its environment, which is an important signifier of its role in woodland conservation

The Walloon branch of the Belgian government agency Matériaux Forestiers de Reproduction has an alien presence in the woodland landscape arising from its simple geometric form and the uncluttered flush surface of the exterior

Moshe Safdie and Associates

The expansion of the Peabody Essex Museum is one of several recent projects by Moshe Safdie that play games juxtaposing animal form and architectural memory. The museum is one of the oldest in the United States, housing an eclectic collection of art, architecture and design from around the world. Safdie's wing provides a focal public space in the campus of existing buildings as well as new galleries arranged as a series of archetypal Federal-style houses – an idea partly inspired by the rows of tombstones in Salem's famous cemetery. These galleries open off a curved, vaulted arcade resembling a fish skeleton. Halfway along this corridor a large glazed atrium projects like the fin of a flying fish.

Safdie admits to a long-held fascination with natural form. It was Louis Kahn's collaborator Ann Tyng who introduced him to D'Arcy Thompson's classic work, *On Growth and Form*, in 1962. He achieved fame with Habitat, a housing megastructure built for the Montreal Expo of 1967 that looked like an efflorescence of some cubic crystalline mineral. Unusually for such an undertaking, Habitat proved so popular that its tenants eventually bought it out of public ownership. The key to its success lay in its manipulation of a modular construction system to produce rich variations.

Although sharing the ideals of Buckminster Fuller, whose geodesic United States Pavilion was the star of the 1967 Expo, Safdie moderates Fuller's utopian vision with domestic familiarity. He accepts that, while science may offer principles applicable to architecture, there are limits to do with the human scale beyond which it can lose its relevance. Even as the designer of the ill-fated Superconducting Supercollider particle accelerator, he suggests that architects who base their work on fractal geometry or quantum physics are using theses sciences merely 'as cover for doing things they are obsessed with'.

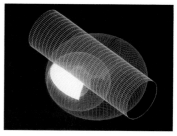

Peabody Essex Museum
Salem, Massachusetts
1996–2003

right: The sinuous, transparent-roofed halls covering the street and piazza spaces provide a beacon for the Peabody Essex Museum as a contemporary cultural focus in the historic centre of Salem

left: Modern biomorphism – concerning forms not directly inspired by nature but generated by overlapping geometric solids – counterpoints the historicism of the galleries. 'The forms are all toroids and quite rationalized, but there is an exciting feeling of being in the gut of something,' says Safdie. 'There is something comforting about that'

bottom right: The bare bones of the glazed spaces recall the wooden shipbuilding tradition of Massachusetts ports as well as making reference to marine skeletons

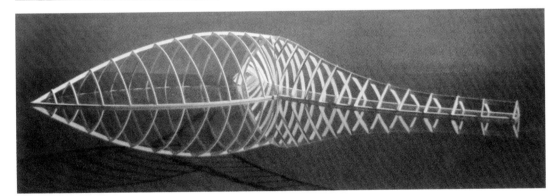

Metropolitan Kansas City Performing Arts Center
Kansas City, Missouri, USA
2000–2007

The outward façade of the Kansas City arts centre is rich in unintended animal allusions, as readers of a local newspaper were quick to point out when it ran a competition to give the project a nickname. The curved segments are non-concentric, but nevertheless assemble themselves to form a kind of city skyline on the entrance side of the complex

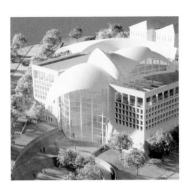

**United States
Institute for Peace**
Washington, DC,
USA
2001–2006

Joining the neo-
classical buildings
along the Mall, the
Institute for Peace
employs purely
cubic and spheri-
cal geometry.
The avian canopy
resonates with
the domes of
the Jefferson
Monument and the
Capitol, but also
sends its own
ambiguous mes-
sage. Is it a dove or
a hawk? Why is it
grounded? One
wing is clearly
fractured – can it
soar at all?

The protective backs that the halls of the Kansas
City Performing Arts Center raise to the outside
world variously resemble armoured creatures from
insects to crustaceans to armadillos, or the alter-
nately narrowing and widening motion of a segmented worm. These were among the similarities
that struck readers of a local newspaper that ran a
competition to find a name for the centre, although
the architect himself had no intention of producing
an animalistic design.

The animal analogy is severely restricted since
the bands or segments of cladding neither lap nor
join smoothly; nor do they continue right across the
body of the building, which metamorphoses unex-
pectedly into a large glazed atrium sheltering the
lobby areas of the centre's ballet and opera house
and concert hall. There is a separate atrium for a
small experimental theatre.

Safdie's initial idea was to evoke various musical instruments by covering the shells in different
materials, but the cost forced a retreat to this more
abstract solution. The segments create a clever illusion: sharing the same radius of curvature but based
on different circular centres, they give the rear of
the building a complex appearance without sacrific-
ing the economy of scale achieved by the use of
repeated components. In other words, the building
envelope does not sweep out a constant silhouette as
if turned on a lathe. This adds interest to the potentially dull blank wall behind each stage. It also makes
all the more surprising the humorous suggestion of a
skyline of a more conventional series of buildings
that suddenly appears on the entrance side of
the centre.

**Yitzhak Rabin
Center for Israel
Studies**
Tel Aviv, Israel
1996–2004

The Rabin Center
is a memorial to
the assassinated
Israeli premier,
containing a
museum, a library,
two auditoria and a
research institute.
The major audito-
rium and the
library project
from the south
range of the build-
ing with glazed
spaces giving
views to Tel Aviv
and the Mediter-
ranean. These are
protected from the
sun by bird's-wing
canopies which
symbolize Rabin's
late conversion to
a belief in achiev-
ing peace with
the Palestinians

Edward Cullinan Architects

Museums around the world compete for visitors these days by commissioning the most modern architecture they can find. This was not exactly an option for the Weald and Downland Museum, which exists to salvage historic buildings in danger of collapse or demolition. The barns and workers' cottages it saves are taken down and reassembled on the museum site. What the museum needed was a space to store these buildings' timbers while its carpenters worked on their restoration.

A farm shed would have done the job, but would not have fulfilled the museum's other wish – to attract visitors to see these craftsmen at work. A bolder architectural statement was needed, but what kind? A new building of a vernacular type would have been all right – but somewhat pointless. False historicism would sit uncomfortably alongside the painstakingly restored exhibits, and nobody does Post-Modern jokes any more. It became clear that a building in apparent harmony with nature would square the circle. The Jerwood Gridshell stands out but seems nonetheless to belong on the site. The novel form, Edward Cullinan found, had the unexpected virtue of immediately placing the project 'beyond discussion' so far as the predictable objections were concerned.

The building aims to be sustainable and uses mainly local materials, but has been realized by sophisticated technological means. The gridshell itself is a square lattice of flexible green oak battens which was assembled flat and then bent into shape to form a roof. The shape is at once complex and satisfying because, although devised on a computer, it is formed naturally as the timbers are stressed. The triple ground-hugging bulges have an insectile quality like the head, thorax and abdomen of a beetle, while the armour of cedar shingles and glazing leads Cullinan to compare the building to an armadillo.

Jerwood Gridshell
Singleton, West
Sussex, England
1996–2002

The cathedral-like
interior, with the
visible texture of
the gridshell
overhead like a
canopied clearing,
renews connec-
tions once made
between Gothic
ecclesiastical
architecture and
the forest. The
Gridshell has a
complex double
curvature for
structural reasons,
but this also adds
visual interest,
giving a landmark
quality to a build-
ing that might
otherwise have
been just a shed or
a barn. Like all
good architecture,
the building has
multiple readings,
with nautical
and aeronautical
as well as
natural allusions

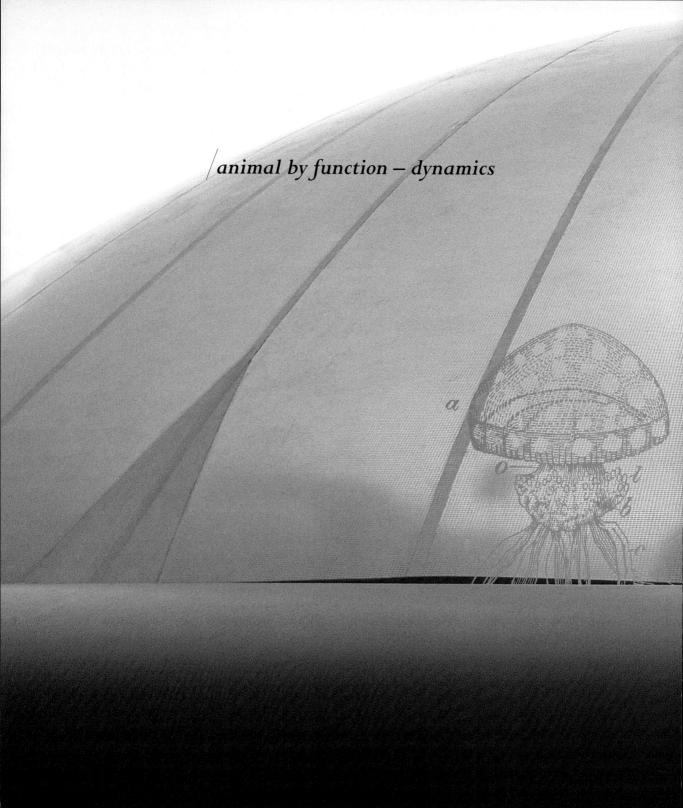

animal by function – dynamics

It seems almost too obvious to state that one of the differences between buildings and animals is that the latter are capable of movement while the former are rooted to the spot. However, buildings do move, albeit slightly, in response to the forces imposed upon them, and the engineering solutions adopted in order to accommodate this necessary movement encourage the emergence of animal form. As buildings incorporate ever more moving parts, used for environmental control and other purposes, it is clear that there are wider lessons to be learnt, not only from the static structures of living organisms but also from the mechanisms by which they and their parts are able to move. Beyond this lie buildings capable of more wholesale movement, a cherished grail of architecture's visionaries, which surely cannot ignore nature's precedent.

The third quarter of the 19th century was the hey-
day of railway architecture. Grand terminal halls
were raised in Europe's great cities. St Pancras
Station in London threw its parabolic steel vault
across a span of 73 metres, the largest in the world.
Elsewhere, engineers crafted train sheds that were
curved and tapered according to the requirements of
their urban settings.

It was not until near the end of the following
century that it became possible to merge these geometric treatments within a single building. Among
British architects, it is probably Nicholas Grimshaw
who most consciously and successfully continues the
tradition of the Victorian engineers. His design for
the international rail terminal at Waterloo celebrates
and updates their achievements – it is London's first
clear-span station since St Pancras – and makes a
fitting terminus for travellers through the Channel
tunnel to France.

The site is tightly constrained by existing railway tracks, roads and surrounding buildings, as well
as by tunnels underneath. Yet the terminal has to
handle 15 million passengers a year. Grimshaw's
solution was to bring the five tracks carrying the
high-speed Eurostar trains in under a long, curving
arch, and to relocate the traditional station concourse below the platforms in order to squeeze
everything into the narrow site. The superstructure
is external to the skin on the west side, where it
supports overlapping panels of glass through which
one can glimpse the River Thames and the Houses of
Parliament; to the east, the structure flips round
onto the inside and is clad in stainless steel.

The tapering asymmetry of the building brings
animal similes to mind. One critic likened it in its
aerial view, with the broad roof of the existing
Waterloo Station alongside, to a caterpillar eating
into a leaf. But the analogy runs deeper than
mere appearance.

Nicholas Grimshaw and
Partners

Waterloo
International
Terminal
London, England
1988–1993

The seductive
convolution of its
light-soaked plat-
form hall makes
the Waterloo
International
Terminal a glory to
rank with the
greatest Victorian
railway architec-
ture. The 400-
metre roof struc-
ture is made up of
36 squashed three-
pin, bowstring
arches spanning
between 35 and
50 metres

Waterloo International Terminal

The west side of
the terminal
building is glazed
with large rectan-
gular panels which
overlap one
another like tiles
in order to fit
the complex curve
of the roof

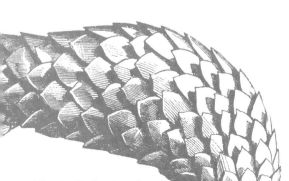

The elaborate superstructure was intended to reflect prestige but not break the bank (it represents in fact a tenth of the cost of the project). This meant controlling the detail design so as to maximize the repetition of parts. The truss configurations were rationalized to just four types, for example, even though the site either curves or tapers over two-thirds of its length.

A system of adjustable supports permitted similar economies of scale in cladding. More than two acres of glazing were made up from large rectangular panels of glass hung from brackets at their upper edge. The other three sides of each panel are free to move with the structure.

These 'scales' make the animal quality of the terminal more pronounced. The comparison is rather strained in visual terms, but is illuminating when other factors, such as the dynamics of the building, are taken into account. As well as providing a fixed shelter over a difficult-shaped site, the steel structure had to accommodate a variety of forces. Along with movement due to the obvious thermal, wind and snow loads, there were settlement effects because the long building rests on a variety of foundations, and, greatest of all, deflection forces due to the mass and deceleration of arriving trains. The freely overlapping glass panels accommodate these movements in all three dimensions, and although the movement is tiny compared with a mobile living creature, the solution reached is strikingly similar.

The project's structural engineer, Tony Hunt, developed a 'design loop' to ensure that the connections permitting this movement would fulfil their purpose. The loop demanded to know: what are the forces; what are the movement requirements; what tolerances are required; is the connection to be concealed or expressed? Such a sequence is implicit in nature's design.

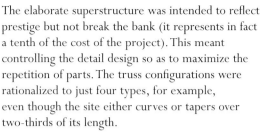

Each glass panel is hung from the steel structure along one edge by means of adjustable brackets, but is left free to move at its other edges, the lower edge being sealed vertically by an accordion-style gasket, while windscreen-wiper-like attachments provide a sliding horizontal seal. The result looks and works rather like the scales of an animal such as the pangolin or the plated lizard

From the air, Grimshaw's new international terminal sidles up to the 1922 domestic mainline station like 'a caterpillar eating into a leaf', according to one critic

The surface of the
Eden biomes is
made up of lentil-
shaped hexagonal
bubbles of ETFE,
a polymer that
admits a broader
spectrum of
natural light than
glass, facilitating
the growth of the
plants within

With its contem-
porary message
of environmental
awareness, the
Eden Project has
revitalized a part
of Cornwall
scarred by disused
clay-pits and
industrial decline

There is no requirement for a building that aspires to a high standard of environmental sustainability to signal this fact by looking like a natural organism. Indeed, one might begin to suspect an architect's motives in adopting such rhetoric. But an exception arises when a building's function is explicitly an environmental one, as at the Eden Project, a visitor attraction on the theme of biodiversity located in a disused Cornish clay-pit. The centre has already proven a stupendous success for its owners, for the region, and not least for the architects, who now find themselves besieged with requests for buildings on similar themes and in similarly appropriate styles.

Nicholas Grimshaw and Partners got the job because of their work at Waterloo, and at first it was envisaged that a similarly sinuous glazed hall might be used to enclose a side of the clay-pit. But it soon became apparent that a series of ultra-lightweight domes would be more suitable in the difficult terrain. Architecture buffs will detect the homage to Buckminster Fuller, and some of them may also identify a debt to Frei Otto, whose experiments with soap bubbles, in the days before powerful digital computers, showed how spheres could be clustered to fit non-circular sites with maximum structural economy. But the hexagonal-framed 'biomes' – the largest 65 metres in radius – that house temperate and tropical botanic gardens have a clear natural precedent, too.

Radiolarians, a genus of marine protozoa celebrated by D'Arcy Thompson, have siliceous mesh skeletons in a dazzling array of polyhedral forms. In the weightless underwater environment, the delicate structures are sufficient to protect the soft body of the creature inside. The Eden biomes, on the other hand, win their battle with gravity because the polymer-film modules cladding the structure weigh just a hundredth of the equivalent glass panels.

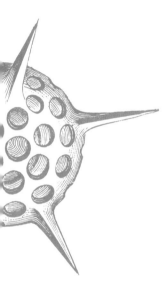

left
The Eden Project

Like the geodesic
domes of
Buckminster
Fuller, the Eden
domes aim to
enclose the great-
est volume for the
least weight – the
largest biome in
fact weighs less
than the air it
encloses. A rigid
structure is
ensured by truss-
like links between
the inner and outer
shell frames

right
**Ludwig Erhard
Haus**
Berlin, Germany
1991–1998

The Ludwig Erhard
Haus is a meeting
point for the econo-
mic community of
Berlin that was
commissioned by
the city's Chamber
of Commerce. Its
ribcage of 15 ellip-
tical steel arches
of different spans
allows floors of
offices to be sus-
pended around a
full-height atrium
and gives the
building its nick-
name: the
armadillo

Airtecture
1996

Airtecture is held firm by a combination of pressurized and evacuated membrane chambers and 'fluidic muscles' which act as tension cables. Using 20th-century technology, its flying buttresses reavow Viollet-le-Duc's famous exaltation of Gothic architecture because it finds precedence in natural forms. Unlike many existing inflated structures, the Airtecture hall provides a conventional rectangular space for travelling exhibitions while also serving as an advertisement for Festo's innovation in materials and devices

Architects have dreamt of using air as a formative element in their work for a century or more, but have been defeated by technical and practical difficulties. Now, however, new fabrics and membranes and improved skill in calculating the forces involved in these radical structures are beginning to see them transported from the realm of fantasy to cost-effective everyday applications.

Festo

Airtecture is the name given to a temporary exhibition hall designed by Axel Thallemer and Rosemarie Wagner, engineers at Festo, a German manufacturer of pneumatic automation components. The structure is made up of 20 inflated roof beams supported by pairs of Y-shaped freestanding columns. These pneumatic flying buttresses are pressurized to withstand the compressive force of the weight of the building. Long, thin pneumatic actuators known as 'fluidic muscles' run both vertically and at angles to the Y-columns, providing the countering tensile force to hold the structure steady. The roof is stabilized both by the pressure in the beams and a partial vacuum in the membrane envelopes between them. In all, the Airtecture hall has more than 300 air-supported elements whose pressure is monitored by sensors and controlled by valve terminals. Inflated, the hall yields a floor area of 375 square metres for an overall weight of just six tonnes, while deflated it can be transported in a standard shipping container.

'It is not by coincidence that many manmade pneumatic structures resemble biological forms,' wrote the German engineer Frei Otto in *Pneumatic Structures*, the first volume of his seminal two-volume work *Tensile Structures*. Thallemer likens the structure to the wing veins of dragonflies, whose hexagonal mesh can be regarded as being made up of repeated Y units. As the imago of the dragonfly emerges from the pupa, fluid pumps through these veins, enabling the wings to unfold for the first time.

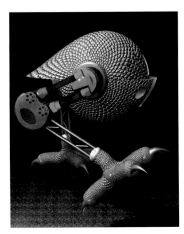

The 'fluidic muscle' is essentially a hose made of high-performance fibre and elastomer layers arranged such that, when the pressure of fluid within it is increased, the hose grows shorter, thereby producing a tensile force. The two 'muscles' here enable a walking action in this study model. Festo's designers examined the arrangement of the highly elastic muscle fibres in worms while developing the device

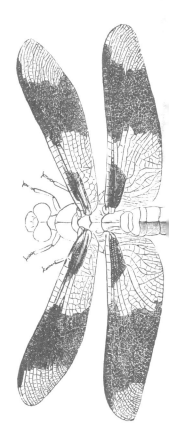

Festo has made the corporate decision to communicate the existence of technologies that might otherwise pass unnoticed by all but specialist engineers to a broad audience through a series of bold vehicular and architectural concepts that use air as a key structural or functional component. These often huge inflated creations are a means of making visible the tiny but technologically highly sophisticated gadgets which are what Festo chiefly develops and sells. The corporate strategy is designed to encourage people to associate the Festo name with a high level of innovation.

Unique creations such as Airtecture make regular appearances at the Hanover Fair and other major international exhibitions, while a fleet of balloons and airships serve as travelling billboards for the company. As Frei Otto leads us to expect, many of these designs inevitably bear a close resemblance to natural forms because of the way they deal with forces, while in other cases nature has provided direct inspiration. Thus, the Funnbrella – a ground-based structure whose name amply describes its purpose – resembles a giant milkcap mushroom, while the form of the Airfish dirigible derives from the shape of a penguin as it swims.

Other enclosed structures of various sizes also cite natural precedents in both form and function. The giant Airquarium is an inflatable shell 32 metres in diameter anchored to the ground by a water-filled perimeter tube. At the other end of the scale, Cocoon is a prototype of a lightweight inflatable tent intended for leisure use or deployment in disaster situations.

More ambitious is Pneumatrix, a portable inflatable auditorium designed by Judit Kimpian, now at Foster and Partners, as a thesis project with sponsorship from Festo. Biochemical and biomimetic ideas inform the organism-like design which, developed with computer assistance, breaks away from the symmetry of earlier Festo structures.

Airquarium
2000

As an aqueous medium forms the 'jelly' sandwiched between cell walls in the bell-like body of a jellyfish in water, so the air in Airquarium's spherical shell between the inner and outer membranes provides its rigidity in air (see also pp.112-113)

Pneumatrix
1997–2001 (concept)

Judit Kimpian's Pneumatrix is intended to explore the potential for pressurized air as a 'smart building material'

Cocoon
1998

Loosely modelled on an insect cocoon, Festo's tent weighs just 1.3 kilograms, with air-filled beams taking the place of cumbersome tent poles

The architecture of international expositions has always called for an exaggerated sense of narrative and technological prowess. Jurij Sadar and Bostjan Vuga's essentially modest Slovenian pavilion proposed for the Hanover Expo packs a punch in both departments. The central feature is an animated tubular membrane covered in polycarbonate tiles which dilates like a snake skin. Within this curvaceous sleeve – in an inversion of the customary architectural dialectic where blobs are contained within boxes – are five cubic exhibition chambers displaying different aspects of Slovenia's natural environment.

Sadar Vuga Arhitekti

A rarity among animal architecture, the pavilion invites visitors into the belly of the beast. The chambers slide back and forth, like bricks swallowed by a snake. As visitors pass from one to the next, they emerge briefly into the 'breathing' membrane and hear recorded sounds of the human body. More important than the passive inspection of exhibits, according to the architects, is the multi-sensory excitement that results as the visitor oscillates between the larger space, with its paradoxically claustrophobic sense of the inner self, and the small chambers with their panoramas of the wider world.

Sadar and Vuga's design for Ljubljana's university sports hall is one of a number of projects for the Slovenian capital that have secured the duo an international reputation since they set up practice in 1996. The objective is to achieve the optimum balance between the enclosed volume, needed for the sports played within, and the available area for seating. With its irregular profile, the 'squashed container' that emerged from this analysis is suggestive not of a carapaced animal like many covered stadia but of a soft organism such as the flatworm, which must strike a similar balance between containing its organs within a manageable volume and maximizing its body area to absorb oxygen.

University Sports Hall
Ljubljana, Slovenia 1997– (competition scheme)

In aiming to optimize the enclosed surface area within a minimum volume, Sadar and Vuga's proposed sports hall presents an irregular profile from all angles, enhancing the appearance of some soft-bodied, bottom-swimming marine creature

**Expo 2000
Slovenia Pavilion**
Hanover, Germany
1998–1999
(unbuilt)

Visitors are swallowed up by the snake-like body of the Slovenia Pavilion, where they are passed through a sequence of cubic chambers alternating with moments where they find themselves once again in more visceral surroundings

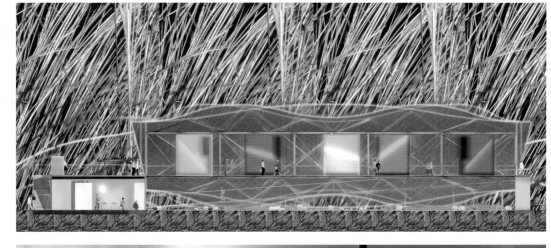

Fountain
Solkan, Slovenia
1998–2001

Like many architectural objects inspired by inanimate nature – the surrounding mountains and rivers in this case – Solkan's concrete-shell fountain also evokes primitive animal forms

D'Arcy Thompson wrote eloquently of the structural bridge and cantilever of an animal's skeleton as it runs from head to tail supported by legs front and back. 'In more ways than one, the quadrupedal bridge is a remarkable one; and perhaps its most remarkable peculiarity is that it is a jointed and flexible bridge, remaining in equilibrium under considerable and sometimes great modifications of its curvature, such as we see, for instance, when a cat humps or flattens her back.' Thompson presented stress diagrams to illustrate the spinal loading of a variety of animals – diagrams that spell out their likeness to manmade bridges.

His diagram of a diplodocus was the starting point for Julia Marks and David Barfield's design for a Bridge of the Future in a competition organized by *New Civil Engineer* magazine. Like the sections of a skeleton, the 200-metre pedestrian bridge is not a traditional arch with fixed feet but a single cantilever. It comprises 23 Y-shaped vertebral units, the largest some 15 metres across, whose size and height is determined by the bending moment at that point in the cantilever. These tetrahedral jacks – based on certain fish vertebrae in fact – carry the compressive load of the structure while steel-cable 'tendons' provide the counteracting tension. Each vertebrum is thus a hinged lever from which the next vertebrum hangs, and the bridge, secured into rock on one side, simply touches down freely on the other with no fixing. A moving walkway runs along the line of the vertebrae.

Movement of the bridge – due to pedestrian load, wind pressure or thermal expansion – is nevertheless minuscule compared with that of animals, and, as Thompson goes on to say, the animal's requirement for great flexibility imposes further difficulties that the structural engineer is thankful to avoid.

Marks Barfield Architects

Bridge of the Future
1988 (competition scheme)

The competition brief for the Bridge of the Future demanded a link with nature as well as a design that would continue the traditional expression of human ambition contained in the great bridges of the past

Eurodrome
Glasgow, Scotland
1990 (unbuilt)

Marks Barfield's proposal illustrates a halfway house between the 'high-tech' ideal of the infinitely extensible shed and organic form. Here, the repeated units do not stretch along perpendicular straight axes, but are bent round circles lying at right angles. The complete shape that results is a toroid, a slice off which forms the building

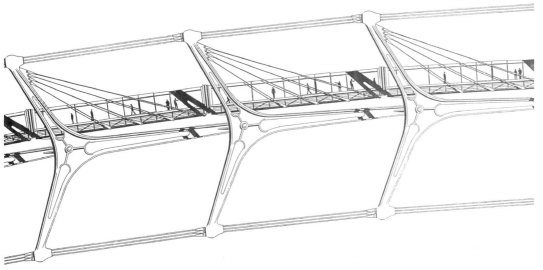

A triangular force diagram of steel or ferrocement 'vertebrae' working in compression, alternating with steel tension cables, provides vertical and lateral stability to the bridge while any torsional moment is opposed by cables running between the tips of the vertebrae and the central axis

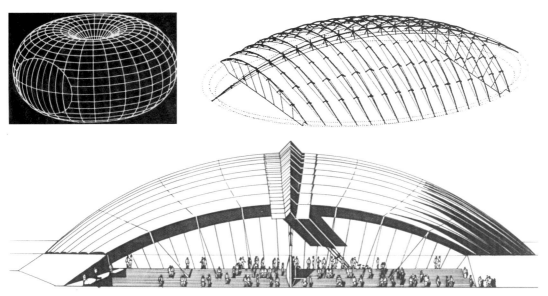

The blister of the Eurodrome is close to the structural ideal recommended by Frederick Kiesler, who, though inspired by nature, rejects the egg in favour of a purer spherical matrix. The shape has made frequent appearances in recent architecture

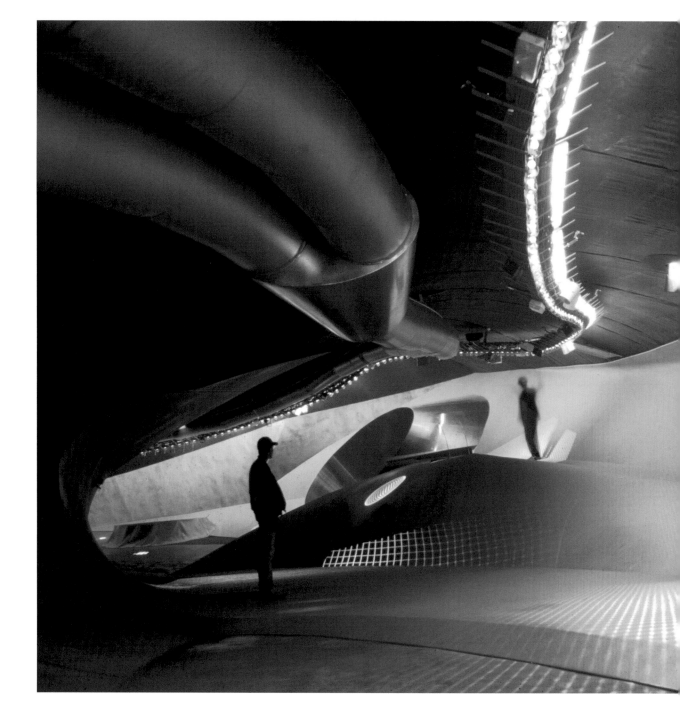

animal by accident

It seems perverse that it should be digital technology that has been so heavily responsible for unleashing the menagerie of continuous, fluid, biomorphic – in other words quintessentially analogue – designs that now dominate the output of younger architects and clamour for attention in international architectural competitions. But in fact it makes perfect sense. For while such designs may start life as a freeform pencil sketch, or a crumple of paper, or a hand-moulded clay model, they must be translated into digital form so that the structural calculations can be performed that will indicate whether or not they will stay up.

Although architects may protest that the computer is 'just a tool', at least one critic has found that close inspection of their work reveals 'the power of the medium to redirect form', and its advent has been responsible for some astonishing transformations in the output of entire offices. Architects now find themselves producing organic designs without any intent to do so – a situation to which some react with equanimity, others with denial. The question, though, is whether this trend is a reflection of the computer tools or the subconscious work of the architects influenced by external cultural currents. (The question is intractable because, of course, the software too may have been designed under the influence of the same currents.)

If this biomorphic architecture is produced by unplanned accident, there are others who wish to enlist computers to produce designs by planned accidents of 'evolution'. According to John Frazer, one of the champions of this idea: 'Architecture is considered as a form of artificial life, subject, like the natural world, to principles of morphogenesis, genetic coding, replication and selection.' The laudable objective is to create buildings that make a better fit with their environment in all senses.

Frazer's 'packet-of-seeds as opposed to the bag-of-bricks approach' promises to take architecture into uncharted waters. However, the biological analogy implied between this and the natural process of evolution is misguided. The correspondences

Asymptote
Fluxspace 1.0
CCAC Institute,
San Francisco,
California, USA
2000

between the environments (built and natural) and fitness (for survival and for purpose) are inexact, and the true Darwinian model (fitness at birth) is muddled with the Lamarckian falsehood (fitness acquired during lifetime).

Architectural design that is the product of so-called genetic algorithms may have much practical merit – it is not merely the conceit of those working with powerful computers and a little knowledge of biology. But just because it is modelled on natural processes of evolution does not mean it automatically arrives at the best answers for building. Biological solutions may be ultimate solutions in their way, but they are also solutions to limited problems, and it is important to be aware of the limits. The dodo was evolution's answer to the task of creating a large bird in an island habitat – and it was only a good answer for so long.

Perhaps the most valid aspect of generative architecture is that it ideally allows environmental parameters to dictate the architectural solution as nature does during ontogeny. Some of the drive in this direction stems from a genuine humility. 'Part of this architecture comes out of negation,' explains Tom Verebes of Ocean D. 'The architectural ortho-doxy is incredibly alien to my generation. We don't want to be like Le Corbusier; we don't want the "end" to be the beginning.' But it is unlikely that many architects will be content to sublimate their egos to the extent of allowing biologically inspired computer systems to do their jobs; they do not like being, to quote the title of Kevin Kelly's seminal work on the coming biomachine age, 'out of control'.

Rotterdam Centraal

Rotterdam,
The Netherlands
2000–2001
(masterplan)
2000–2006 (station)

left: The bulbous forms of the buildings planned to rise above Rotterdam's new transport interchange reflect the need to maximize lettable space without loss of public space for pedestrians at ground level. Individually and collectively, Alsop's buildings bear an unintentional resemblance to the elongated, chopped bulb shape of tunicates, marine invertebrates which gather in colourful seabed colonies. Even the faceted glass cladding finds an echo in the transparent 'tunic' that gives the creatures their name

far left: Similar formal language – derived not in fact from biology but from mid-20th-century art – permeates the interior spaces

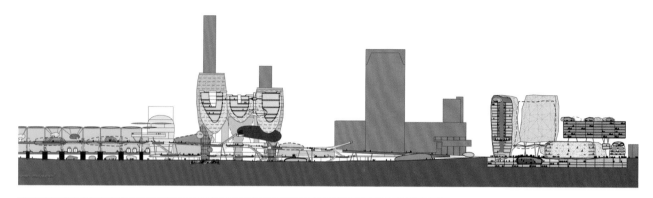

left above:
Uninhibited by any
notion of taste that
says a city can only
stand so many
irregular buildings,
Alsop's interven-
tion on Mauritsweg
(towards the right
in the section),
and in the rest of
central Rotterdam
(to the left), begins
to imagine a
biomorphic urban-
ism, where nature
may inform not
only the morpholo-
gy of individual
buildings but also
the relationship
and the nature of
the spaces
between them

left below:
Irregular shapes
and surfaces
characterize interi-
ors and exteriors
alike of Alsop's
Mauritsweg rede-
velopment, which
includes retail,
cultural, commer-
cial and residential
space

Will Alsop's scheme for the redevelopment of
Rotterdam central station and some of the major
surrounding streets provides one of the most
dramatic proposals ever conceived for a biomorphic
cityscape. However, biological models played no
conscious part in Alsop's thinking when he devised
the plans.

The 20-hectare development is planned to
include 700 new homes and 200,000 square metres
of office accommodation as well as retail and cultural facilities. The focal point is the transport hub,
which provides an interchange for trains, metro and
buses, and parking for cars and bicycles. Rising
directly above it, and strung out along the avenues
that run either side of the railways tracks, are wild
clusters of bulbous forms making a new business
axis for the city.

Elements of Alsop's masterplan will be seen
first, however, along the perpendicular cultural axis
that stretches south from the station along the
Kruisplein and the Mauritsweg. Plans here involve
the redevelopment of a 1960s city block containing
a church, a hotel, and shops. These 'rocks' are to be
remoulded in a more amorphous fashion and joined
by new apartments and offices, all connected by a
two-tier pedestrian circulation route, one route
placed under the transparent pavement of the other.
The 'rocks' and the similar forms over the transport
hub may be read in many ways. Colourfully glazed,
they are like Tiffany vases or crystal goblets; striving
skywards and spreading out to receive the sun, they
are like trees in a forest glade – and perhaps a satire
on the trend for rapacious air-rights developments
above public transport facilities. But for a resemblance in nature of this cluster of rising blobs, it
would be hard to better the marine invertebrate
chordates known as tunicates which form brightly
coloured colonies on the sea floor.

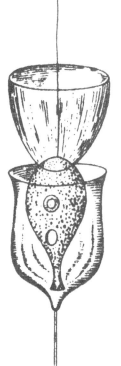

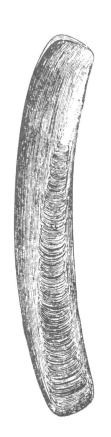

Le Grand Bleu, as the headquarters of the French Département of the Bouches du Rhône has become known, amply illustrates Will Alsop's artistic approach to architecture. It makes a startling contrast with the bleached appearance of local buildings and the rock they are built upon, and provides a bold symbol of civic commitment in a depressed suburb of Marseille. The site, on an exposed hillside, demanded that special consideration be given to wind loads, including that of the ferocious mistral. This and other environmental factors, such as the wish to rely as much as possible on natural ventilation, influenced both the external form and the internal layout.

The main body of the building comprises two long blocks of offices on either side of a top-lit atrium. This rectangular public space has a sloping floor, which gives it an outdoors feeling, and is inhabited by a number of more sculptural interventions containing visitor facilities. Still more dramatic are two large free forms clasped to the top and side of the main building. A tubular form with a squashed, aerofoil section lies like a razor clam along the full length of the roof, and contains the offices of senior dignitaries. To the side of the building, and reached by means of a number of covered bridges, is the more sculptural and geometrically complex Déliberatif, the council chamber and assembly hall. This too is an aerofoil in section, but it is also tapered and, with its fabric sunshade stretched over its structure, rather resembles a chrysalis.

Hôtel du Département des Bouches du Rhône
(Alsop & Störmer)
Marseille, France
1990–1994

The rectangular, tubular and ellipsoidal forms that make up Le Grand Bleu illustrate evolutionary stages in the geometry of structures. The contrasting shapes are clustered to minimize the effect of high winds and are unified in visual terms by the overall bold blue finishes. The complex envelope of the insectile Déliberatif is clad in triangular glass panels. Both form and colour provide interest for this prominent landmark building

Cardiff Bay Visitor Centre
(Alsop & Störmer)
Cardiff, Wales
1990

The Cardiff Bay Visitor Centre was designed as a low-cost, temporary structure to house exhibitions on the redevelopment of the area. Prototyping forms seen in later Alsop buildings, it nevertheless illustrates its architects' ability to use memorable form and technical innovation to create a popular new architecture of place. A track-like steel substructure supports a series of elliptical steel ribs which are covered in plywood panels and a weatherproof fabric skin. The ends are glazed

Stonebridge Health and Community Centre
London, England
2002–

Four modules are uniformly clad in enamel panels of varied shape and colour developed in painting workshops with the local community. Such panels provide an elegant and increasingly popular means of finishing amorphous buildings where it is impossible to finesse the structure or the weatherproofing to aesthetic satisfaction

Dental School, Queen Mary & Westfield College
London, England
2000–2003

Alsop's blob-like interventions find a literal rationale in the dental school and research facility on the Whitechapel campus of Queen Mary & Westfield College in London. Brightly coloured forms like biological cells or molecular models suspended within the glass atrium house seminar rooms and some offices, and lend the building an appealing personality for its users as well as passers-by. As is the case with many new research centres, the brief called for an open environment that would promote creative encounters between the research staff in different departments. The glass envelope provides the desired transparency, making each department readily identifiable within, while open spaces within provide areas where staff and students may gather

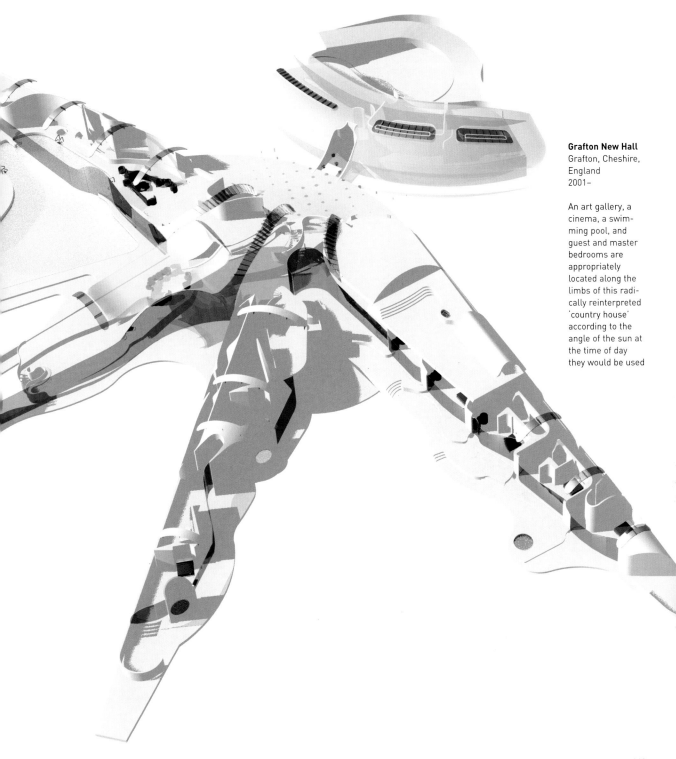

Grafton New Hall
Grafton, Cheshire,
England
2001–

An art gallery, a
cinema, a swim-
ming pool, and
guest and master
bedrooms are
appropriately
located along the
limbs of this radi-
cally reinterpreted
'country house'
according to the
angle of the sun at
the time of day
they would be used

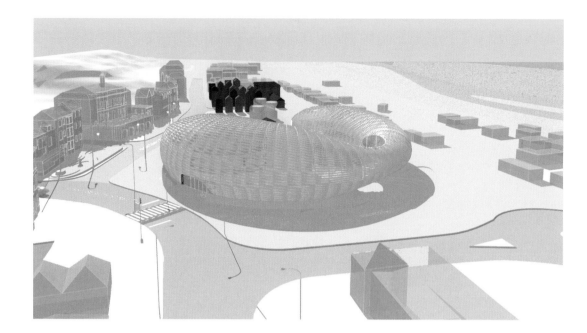

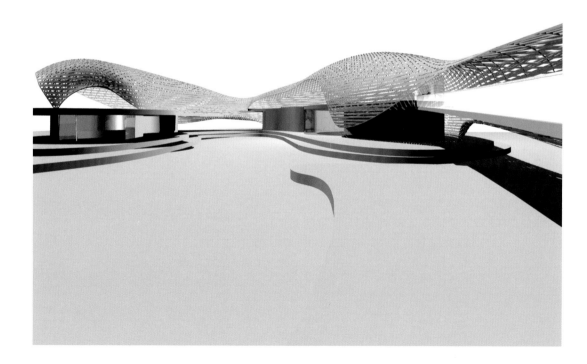

The Stade aims to
blend in with its
surroundings not
by mimicking the
style of town build-
ings nearby, but in
a more fundamen-
tal sense by allow-
ing the local flow
of people to play a
role in form-giving.
Shiplap timber
construction
echoes the
distinctive black
clapboard towers
where Hastings
fishermen
hang their nets.
The 'head' section
of the Stade
houses a restau-
rant, while the
body accommo-
dates an informa-
tion centre and
exhibition space. A
viewing terrace is
situated on the
neck that joins
these two sections
of the building, and
visitors exit
through the 'tail'

For Kathryn Findlay, the form of a building arises
from 'the invisible things that generate shape' – the
flow of vehicular traffic around, and of pedestrians
around or through, it; solar and thermal considerations. Of course, these issues are traditionally considered – or supposed to be – by architects whose
work looks quite conventional, so there must be
other influences at work in a scheme such as Ushida
Findlay's Stade Maritime.

Located on the seafront at Hastings, the Stade
is a visitor centre and mixed-use development that
aims to provide a new focal point in an economically
depressed town. Ushida Findlay's competition-
winning proposal makes a bold formal contrast with
its largely Victorian surroundings, but echoes the
local fishermen's shacks and boat-building tech-
niques by using shiplap timber cladding.

In what is still an important commercial fishing
centre, the building curls round a courtyard like a
freshly landed catch or one of the lugworms fishermen use for bait. In the competition proposal, the
body of the building was oriented towards the town,
but after consultation the plan was flipped so that
the courtyard opens on the town side. The building
rests lightly on the site, like a boat pulled up on the
beach, with room to pass underneath.

Like Snøhetta's cetacean Turner Centre at
Margate and Birds Portchmouth Russum's more fig-
urative shrimps at Morecambe, it has some kinship
with the surreal marine creatures designed for the
1992 Barcelona Olympic Games by Frank Gehry
and the local designer Javier Mariscal (which in turn
may derive from the prevalence of crustaceans in
Spanish surrealism). All these buildings are
reminders that strange rules apply to seaside architecture.

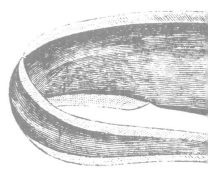

Grafton New Hall
Grafton, Cheshire,
England
2001–

Grafton New Hall
is intended to
appear as a low
sandstone outcrop,
with grass growing
up its sides as if it
has been there for
centuries. Despite
its large surface
area, the form and
orientation of the
house and its par-
tial burial in the
landscape mitigate
winter heat loss
and summer solar
gain

The English country house has been in sharp decline since the Edwardian era. However, a single affluent client may be all it takes to revive the concept for the 21st century. Grafton New Hall is not the throwback one might expect, but a deep recon- sideration of the idea of the country house taking into account contemporary social and environmental sensitivities.

Ushida Findlay find an unlikely strand of inno- vation in this building type. 'There is a strong desire to use the country house as a laboratory for new design approaches,' Kathryn Findlay suggests. Examples such as the glazing at Hardwick Hall, Paxton's irrigation at Chatsworth, Prince Albert's recycling of organic waste at Osborne, and William Armstrong's employment of hydroelectric power at Cragside confirm the truth of this initially surprising observation.

Grafton takes its place in this lineage, but dispenses with the pompous appearance that tradi- tionally signifies dominion over the land and the local peasantry. Its low profile makes it, in the archi- tects' words, 'an iconic building that redresses the country house typology as an expression of status. Instead it is stealthy, a subtle building where the transition between the architecture and the land- scape is seamless.'

Starfish limbs radiate enough like the wings of a traditional house. And, like a starfish, the house clings to the land. The visual symmetry of the build- ing envelope does not reflect the internal arrange- ment as it does in the biological analogue, however. Various functions – bedrooms, pool, a cinema and so on – are disposed along the limbs according to a logic given by the track of the sun and the times when the spaces are most likely to be used. The occupants thus move from room to room during the course of the day – much as they would have done in many a traditional country house.

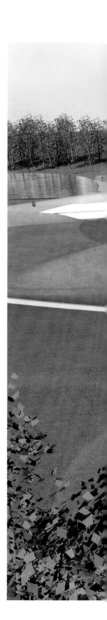

The Truss Wall House, Ushida Findlay's second ever commission, arose from the architects' interest in a novel construction system, the 'truss wall', which allows unusual shapes to be formed in reinforced concrete without loss of structural integrity. The system proved too expensive for use as intended in low-cost construction projects, and by the time that Ushida Findlay discovered it, it had found a less satisfactory outlet in the manufacture of giant Buddhas and cartoon characters.

The architects' unconventional approach to topology gave new scope for an exploration of the system, which divides walls into vertical 20-centimetre sections. Steel reinforcement is bent at these intervals to follow a desired shape before wire mesh is laid and concrete poured over. The extreme flexibility of the system gave freedom to derive the plan of the house from the routes that its occupants might take through it. This led not to standard walls and floors but to 'pliable viscera', which were in effect frozen into shape as they fell into the envelope of the house.

The old idea of a formal sequence of static, proportioned spaces was jettisoned in favour of a sense of constant flow. Ushida Findlay wrote of the house as 'translating the orthodox architectural language of modularized floors, walls and ceiling into a kind of slimy fluidity'.

The feeling that there were botched details and wasted space lurking in the irregular interstices behind the smoothly curving cement skins would doubtless offend a designer wedded to the Cartesian grid, but Kathryn Findlay is unapologetic. There is no need to resolve these boundaries and junctions as long as they are not in the way. Flow is everything.

Soft and Hairy House
Ibaraki, Japan
1992–1993

This house is Ushida Findlay's response to Salvador Dalí's demand that 'future architecture should be soft and hairy'. The plan is loosely anthropomorphic, using the client family as models, while the bathroom bubbles out into a central courtyard like a coral polyp

Rooftop Garden, Tate Modern
London, England
2001 (concept)

Ushida Findlay envisioned a rumpled metal mesh distended where fastened potted trees weigh it down and altered in shape to form 'a living landscape' as it responds to both instantaneous and gradual changes in load from wind in the trees to the gathering weight of foliage

Truss Wall House
Tokyo, Japan
1990–1993

The Truss Wall House makes use of a flexible concrete construction system in order to re-evaluate the topology of the home based on the ideal pattern of movement of the occupants through it. The fluidity of the interiors is enhanced by the chosen finishes. This courtyard floor (below), for example, is made of liquid mortar poured into balloons and then packed to form a hexagonal array like that seen in tortoiseshell and in many other instances in nature

Embryological House
1998–2000
(concept)

On one level, the Embryological House represents an attempt to give people individual solutions within a mass-production market system. But on another, it is a celebration of computer design or, as Lynn puts it, 'an unapologetic investment in the contemporary beauty and voluptuous aesthetics of undulating surfaces'. Mindful of the needs of a developer to market the concept as a brand, Lynn has ensured that the houses share certain properties – each has two storeys, with similar floor areas, and the upper floor making a gently sloped living area – as well as a family resemblance. Nevertheless, each design is unique

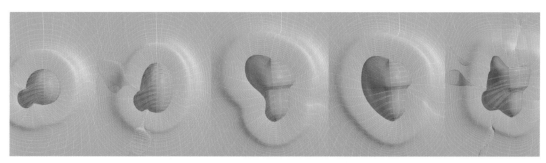

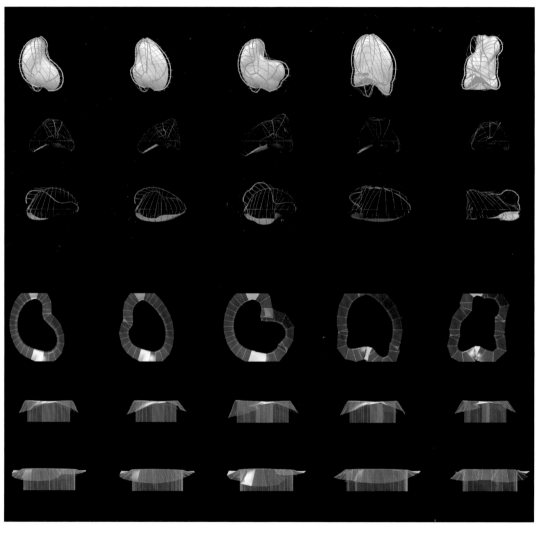

In a good-humoured running debate, Frank Gehry gently chides those who claim to use the computer to generate their architectural forms. The image in the head ultimately takes precedence over the image on the screen, he insists, even though computer users might believe themselves to have generated the form extracorporeally. One of Gehry's chief antagonists must be fellow Los Angeleno Greg Lynn, who believes that the computer tools now available give architects cause to set aside history and theory and all the other stuff that clutters the insides of their heads.

Lynn's demonstration piece is the Embryological House, a series of house designs varied for individual clients and locations using Maya software. Each house is constrained within certain geometric parameters – for example, they are all two storeys, with the upper-storey living area sculpted into a dish shape. But once these limits have been set, infinite variation is possible in other aspects of the design. In this case, variable quantities such as local conditions of climate, the site and building materials as well as functional and aesthetic requirements of each client are manipulated by the computer to influence the form.

What results, perhaps not surprisingly, are naturalistic forms a world away from the box-like repetitiveness of standard mass-produced houses. In order to make this departure, Lynn looked at symmetry-breaking during epigenesis, when a fertilized egg divides into two cells, then into two more and so on, losing symmetry as it goes. The houses are not meant to look like embryos – even if the inchoate shapes that the computer throws out do suggest this image; it is the embryological *process* that is the architect's guide. 'This marks a shift from a Modernist mechanical kit-of-parts design and construction technique to a more vital, evolving, biological model,' says Lynn.

Greg Lynn

Six prototypes of the Embryological House were fabricated in plywood, illustrating the range of constraints – spatial, functional and aesthetic – that determine both the common and the varied aspects of their form. As Lynn points out, 'there is no ideal or original Embryological House as every instance is perfect in its mutations'

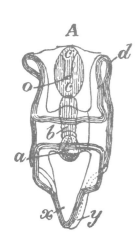

Lynn warns that computers risk tempting architects into superficial mannerism. In his book *Animate Form*, he writes: 'The challenge for contemporary architectural theory and design is to try to understand the appearance of these tools in a more sophisticated way than as simply a new set of shapes.' The challenge is very real, because the qualities of flow, motion, and change over time that computers enable architects to bring into consideration are precisely the qualities that they have hitherto been forced to neglect as they manipulate static form in static space. To drive home his point, Lynn compares the revolution not with any stylistic development, but with the technical development of stereometric projection and perspective during the Renaissance. The most important thing to realize, he says, is that 'these technologies are animate'. Just as the understanding of perspective enhanced the development of architecture in the three dimensions of space, so the computer now promises to introduce the fourth dimension – time.

At the root of Lynn's biologically inspired remedy lies the belief that, as Ernst Haeckel and others succinctly put it, 'ontogeny recapitulates phylogeny' – in other words, that the stages of development of an organism replay the evolution of the species at high speed. This notion has long been discredited in biology, but its analogue in architecture casts a powerful spell, suggesting that if buildings can be 'evolved' within the space of a single building cycle thanks to the computer, then answers to the old questions of commodity, firmness and delight that have taken centuries to reach (if we've reached them yet) might be arrived at almost immediately.

Arca del Mundo, a
tourist attraction
on the theme of
ecology commis-
sioned by the
Costa Rica govern-
ment, understand-
ably takes a more
explicit biological
form than most of
Lynn's projects.
The architect has
suspended his
usual rules for the
'evolution' of form
and allowed him-
self to produce a
design inspired by
local plants and
animals. The main
body of the build-
ing comprises a
series of bulbous
chambers for exhi-
bitions of contem-
porary art and
local culture
arranged around a
central space,
from which a heli-
cal stairway rises
to a foliate fabric
canopy that pro-
vides a vantage
point to the
surrounding rain-
forest treetops.
Ecological subject
matter is displayed
in the ground-floor
'Consilience
Museum', based
on the theories of
the naturalist
Edward O. Wilson

Kunstmuseum
St Gallen,
Switzerland
2001 (competition
scheme)

The St Gallen
Kunstmuseum is
home to collec-
tions of both nat-
ural history and
contemporary art,
making it an ideal
client for an archi-
tect like Greg Lynn.
An underground
passage links the
existing building to
an extension which
hoists archive and
storage spaces
aloft on three fluid
volumes containing
new sculpture gal-
leries. A complex
mesh of tendrils
provides the arma-
ture for the three
special galleries
and contrasts
effectively with
the more conven-
tional steel-frame
structure above
and below

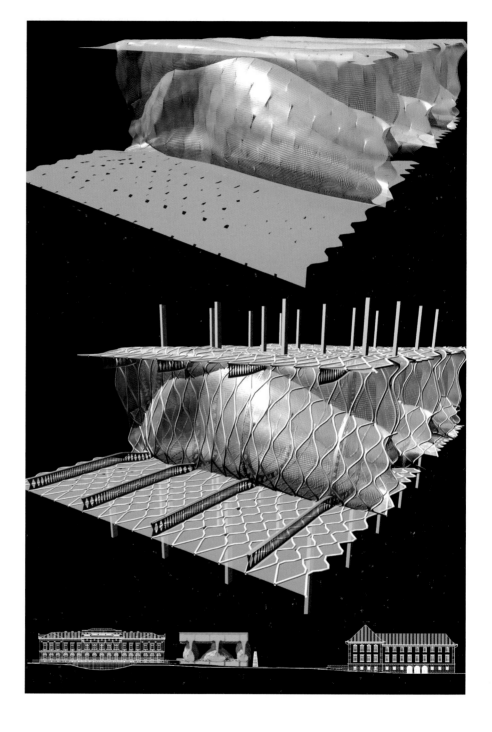

The 'animate' properties of computer-generated form do not of course require architecture to be modelled on life forms and life processes. But the biological paradigm seems obvious in the circumstances, and it is only natural, as it were, that architects should look here first for guidance as they begin to explore the creative possibilities of the computer. This scenario raises the possibility that the present eruption of biomorphism is in fact just a first pass in what could yet prove to be an unimaginably rich (visual and functional) new architecture that will only be realized when the potential of computers to contribute to the generation of built form is more fully understood.

Against this optimistic vision of the technocrat, however, it is important to set a more realistic view and remember that domestic architecture especially is governed by tremendous conservatism. Greg Lynn has been described as bringing Buckminster Fuller's Dymaxion House into the era of genetic engineering. But the Dymaxion House never caught on with the public, or much influenced domestic architecture. Whether the work of Lynn and his contemporaries finds greater favour remains to be seen.

Uniserve Corporate Headquarters
Los Angeles, California, USA
2001–2002

An important feature of computer-generated architecture, also found in biological precedents, is an ability to scale details in a 'fractal'-like way. Lynn's partition walls for a corporate conference room, an elongated plastic enclosure inserted into the loft-like office space, has a texture that varies according to the local geometry like elephant hide

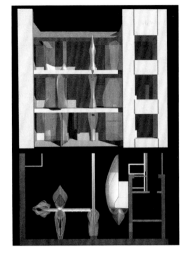

Imaginary Forces Offices
New York City, New York, USA
2000–

Lynn collaborated with the film-titles design company Imaginary Forces on two-dimensional projects before being asked to design its offices in Manhattan. Lynn slices through the three storeys of the building that the firm occupies with translucent fibreglass partition walls which are pierced and pockmarked with further partitions, shelving or light wells. These blisters or blebs occur like epigenetic events along the otherwise continuous surface

The liquefaction of existence demands a liquid architecture. As our identities become more fluid, so must our environments. In particular, the grid, the organizing principle for traditional architecture, is obsolete: imposing order by imposing a grid upon an area is a top-down, implicitly domineering action ill-suited to creating buildings that involve their users more deeply. For Lars Spuybroek of NOX, who makes this argument, computers provide a remedy; they are able to generate architecture not merely imitative of natural fluids but, like liquids, and like our own increasingly mediated sense of ourselves, ever liable to transformation.

It seems odd at first that this radical agenda should produce architecture susceptible in particular to biomorphic decoding. 'Fluid' and 'organic' are not the same – after all, a liquid under gravity finds its own level in the pure and simple geometry of the flat plane. Perhaps we interpret 'fluid' form as organic because we seek out natural analogies, or perhaps the organicism is put in as an artefact of the software. And yet, of course, the form of an organism is also determined in proportion to the resistance it encounters in the fluid medium, air or water, that supports it.

Spuybroek believes analogously that the electronic media we swim through should reshape our architecture. In 1994, NOX received a gift of a commission to demonstrate its ideas. The government client required solid architecture and virtual presentations to work together in an exhibition pavilion aimed at educating the public about the importance of water. The plywood-covered steel structure of the 'WaterLand' pavilion forms 'a braid of splines', which allows walls to flow into floors and, assisted by real and virtual water features, promotes a sense of the visitor's merging with water. The result is an experiential installation where architecture becomes a psychophysical extension of the person.

Beachness
Noordwijk, The Netherlands
1997 (concept)

A boulevard and beach hotel blends landscape and seascape, natural and designed space, real and virtual environments. The hotel is a cluster of bubble capsules, like sand and water being thrown up into the air

Pavilion for the Delta Expo 'WaterLand'
Neeltje Jans, Zeeland, The Netherlands
1994–1997

NOX's freshH$_2$O eXPO pavilion may resemble a washed-up fish or a seal having a scratch; however, any visual likeness to actual life-forms is not the point, more an accidental side-effect of the wish to produce fluid forms better suited to the nature of contemporary life. Spuybroek is drawn to the phantom-limb phenomenon felt by amputees, and is interested in producing similar alterations to the human body's self-perception by combining architectural and media spaces (see pp. 130–131)

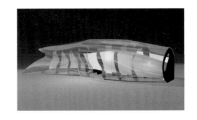

**Blow Out toilet
block**
Neeltje Jans,
Zeeland, The
Netherlands
1997

The seal's pup?
Similar computer-
generated spline
geometry is used
for a toilet block
lying near to the
water pavilion. The
open-ended struc-
ture enlists the
wind as yet anoth-
er of the 'media' by
which sensory
experience may be
heightened even in
the most function-
ally driven of archi-
tectural projects

Combining offices with children's play space and retail facilities, the SoftOffice headquarters for a children's broadcasting company set a demanding target in NOX's mission to dissolve traditional boundaries in architectural planning. The children's space is based around objects which are the vehicles for real and internet-based interactive play. A focal point is the 'Glob', a physical and internet environment designed by globally networked children, which behaves in some senses like a 'living organism', with genetic algorithms used in programming its interactivity. In this lively environment, Spuybroek is amused to find, waiting parents become the most architectural features of the space, standing idly around like caryatids.

The office space seeks to address the increasing problem that, while spaces designed for single functions may focus attention on that function but are becoming outmoded, the supposedly multifunctional spaces that would replace them are inadequately conceived and may simply lead to dissipation of their occupants' efforts. Spuybroek aims to replace the 'passive flexibility of neutrality with an active flexibility of vagueness'. What this means in practice is that the uncertain pattern of work teams' grouping and regrouping is reflected in an architecture of apparently random expanded and contracted spaces for meetings or concentrated individual work in contrast to the conventional office with its personal territories.

The spaces were modelled in a manner reminiscent of Frei Otto's soap-bubble designs, using a highly flexible mesh of slack rubber tubes fastened across two wooden rings that were then dipped in lacquer and pulled apart. The preferred complex three-dimensional lattice that emerged was then manually transcribed into digital form for later realization in timber construction.

wetGRID installation

'Vision Machine' exhibition, Musée des Beaux Arts Nantes, France 1999–2000

An exhibition installation in a classical courtyard shows Spuybroek's treatment of the grid. The lines of the grid are not broken, but in a 'self-choreographing' procedure are crossed and forked and spliced into a complex net by the action of 'vortex forces' to produce a set of tentacular volumes that inescapably suggest formations in nature

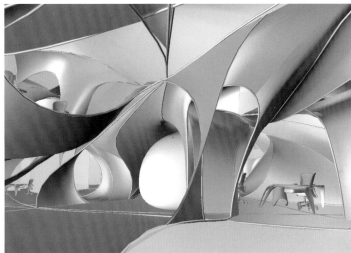

SoftOffice
England
2000–2005

The SoftOffice project for a children's broadcasting company brings office space, retail space and a children's play area together under one roof. The interiors of the SoftOffice were modelled in the first place by analogue means by dipping webs of rubber into lacquer and then pulling them apart as the lacquer dried. The arrangements produced have structural integrity by virtue of the process as well as a natural precedent in the cancellous tissue of porous bone

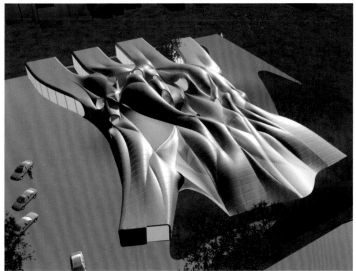

Cardiff Bay Opera House

Cardiff, Wales
1994 (competition scheme)

The use of computers is leading architects such as Reiser + Umemoto to reassess the causal relationship between form and function. The echoes of Constructivism and the animalism of the resulting projects are logical, if perhaps surprising, consequences. Here the bodies of the theatres are cantilevered from the rectilinear fly tower and supported on long struts, endowing the scheme with an insectile personality

Yokohama Port Terminal

Yokohama, Japan
1995 (competition scheme)

Reiser + Umemoto's proposed port terminal is topologically similar in structure to many transport buildings but is a radical departure in morphological terms, a transformation underlined by the fact that the building's roof is intended to be landscaped as a park

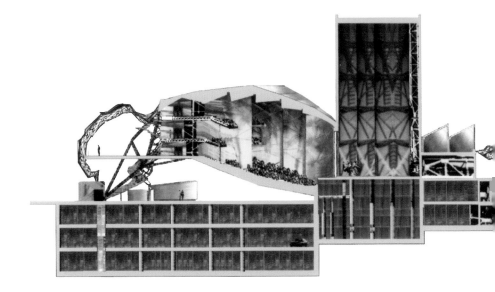

The recent portfolio of Jesse Reiser and Nanako Umemoto neatly illustrates the progressive shift now under way from an industrial aesthetic towards a more adaptive, mutational style of building. One focus of their attention has been the space frame, the lightweight three-dimensional framework that is many architects' preferred solution for enclosing large spaces. As these spaces – typically, convention centres, exhibition halls and transport interchanges – have become ever more bloated, so the realization has grown that beyond a certain size they lose the ability to fit into the built landscape. The effect is graphically – and indeed geographically – illustrated by Foster's Chep Lap Kok airport at Hong Kong and Piano's at Kansai, both on artificial islands. These structures try to ameliorate their vast scale with sinuous elements but still prize repetition as their primary aesthetic force.

Reiser + Umemoto acknowledge that the supremely economical space frame cannot be rejected entirely, and instead seek to modulate its structural principle. 'It now becomes possible to conceive of the space frame's urbanistic potential within a new paradigm – one of continuous variation,' they write. Although nature excels at producing examples of such varied repetition, the architects draw no explicit biological analogy into their work.

The Yokohama Port Terminal is essentially a long triple-arched shed, but is broken up in scale by the uneven repetition of the structural units of the frame made possible by the use of new CAD tools. This so-called 'amorphic' space frame has been developed in subsequent projects, such as a 1999 proposal to enclose a vast area of New York City's West Side (ironically including the Javits Convention Center, a prime example of monolithic space-frame construction), and in Reiser + Umemoto's entry the following year in the competition to design BMW's new customer reception centre in Munich.

BMW Event and Delivery Centre
Munich, Germany 2001 (competition scheme)

Recent projects loosen the formerly rigid geometry of the space frame

Ocean D

The Modernist ideal, expressed in buildings such as the Centre Pompidou, was to create maximally flexible interior space (never mind that the designed-in flexibility was seldom exploited). One reason for the emergence of this ideal was simply that it was difficult and expensive to plan varied spaces. Now, with the help of computer-aided design and manufacturing, groups such as Ocean D believe it is possible to apply an almost opposite strategy, and offer instead a rich variety of spaces within which different uses can find their place.

A new mood acknowledges the futility of planning for extreme specificity – because as soon as spaces are used, they are misused, and their functionality is blurred with that of adjoining spaces. People adapt to buildings anyway, the thinking runs, and for that matter buildings adapt – in ways still poorly understood – to however they are used.

Ocean D's proposal for Rabin Square in Tel Aviv demonstrates the potency of the blend of determinism and quasi-Darwinian laissez-faire that characterizes the firm's computer-assisted architecture. A sequence of 18 complex looped steel figures, intended to be fabricated under CADCAM control, are strung out across the square, emanating from the point where Yitzhak Rabin was assassinated. The gently rising line drawn through the centres of the figures is a given, along with various parameters of the structural components. These data points combine with free variables to generate the shape of each object and the immediately surrounding topography, which in turn affects the passage of pedestrians through the square. Planting and water pools at points across the square give nature the chance to add to the mix of predictable and unpredictable – puddles will always form and plants grow in the same places, but the weather and the seasons will determine how they appear at a given instant.

**Latent Utopias
Exhibition**
Graz, Austria
2002

An installation of five objects, not alike but similar in their 'genetic code', runs alongside the 19th-century Landesmuseum Joanneum exemplifying 'a manufacturing logic that affords continuous serial variation'. The volume of each object and the number of steel tube splines and welds from which it is made up are 'scripted' elements of the design; other variables are set within ranges that give the computer scope to generate forms with surprising characteristics

**Rabin Peace
Forum**
Tel Aviv, Israel
2001 (competition
scheme)

The square in Tel
Aviv outside the
building where
Yitzhak Rabin was
assassinated is the
city's most impor-
tant public space.
Ocean D's plan
combines the func-
tions of a memori-
al, a public park,
and a site for
demonstrations.
The undulating
landscape and the
sculptures that
march across the
square are calcu-
lated in part based
on the conjectured
movement of
pedestrians
through the space

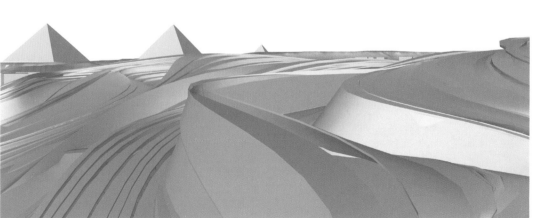

**Great Egyptian
Museum**
Giza, Egypt
2002 (competition
scheme)

The skin of the
building, shaped
as if by natural
forces, illustrates
the potential for
computer algo-
rithms to produce
architecture that
resonates with its
landscape. The
dominating pres-
ence is the roof
whose apparently
irregular topog-
raphy is made up
of a lattice of
modular spline-
curved structural
units

right
HydraPier
Haarlemmermeer,
The Netherlands
2001–2002

The HydraPier is a
municipal pavilion
on a lakeshore in
landscape adjacent
to Amsterdam's
Schiphol Airport. It
is an architectural
gesture reflecting
the contrasts of its
artificial pastoral
setting with jumbo
jets screaming
overhead. Fluid
and organic forms
were clearly
essential elements
in helping to artic-
ulate the locale's
struggle between
land and water,
nature and techno-
logical artifice

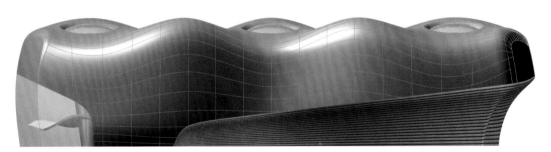

left centre and
below
**Mercedes-Benz
Museum of the
Automobile**
Stuttgart, Germany
2001 (competition
scheme)

Asymptote saw
their task as one of
rendering the inte-
grated 'technologi-
cal precision and
aesthetic sophisti-
cation' of
Mercedes-Benz
cars in built form.
Their response
involved the use of
simple intersecting
wave geometries
to create a seam-
less undulating
'landscape' for the
display of historic
vehicles. A variety
of transparent,
louvred, and
opaque surfaces
are used to clad the
building, providing
surfaces both
inside and outside
that are continuous
in appearance but
variable in their
properties. The roof
is made up eco-
nomically of sine
and cosine wave-
forms, although
the blobby organic
appearance of the
overall envelope
belies this mathe-
matical purity

In the new technological aesthetic, form is everything. Surfaces alone communicate the complexities
– technical, cultural, contextual, historical – of the
work, and structure must do its job as best it can
behind the scenes. For the Mercedes-Benz Museum
of the Automobile, Lise Anne Couture and Hani
Rashid of Asymptote proposed to wrap the exhibition space in a wavy skin. The design has a marked
organic appearance, but organicism was not high on
the architects' agenda. It was merely a secondary
effect: because cars now have organic shapes, and
because Asymptote wanted their building to have 'a
smooth, taut surface not unlike
that of a Mercedes-Benz car body'.
In other words, this updated
machine aesthetic simply happens
to have biological overtones
because this is where CAD has
already led product design and car
design in particular. The structure,
the architects added almost as an afterthought in
their own description of the project, 'follows closely
the architectural form of the building'.

    Formerly, such a statement would have been a
nonsense: of course the structure follows the form
of the building, or rather the other way round. But
there is now the choice. No longer tethered to the
structure, the surfaces of buildings are free to
engage with other narratives. Principal among them
for architects such as Asymptote are the planes of
virtual space. Asymptote believe that the fusion of
real and virtual space will characterize the architecture of the future. What is curious is that the forms
that emerge are so often markedly 'fluid'; and equally curious, though for reasons psychological rather
than technological, is their propensity to suggest
organic analogies. It hardly seems a matter of chance
that Asymptote's most significant exercise in real–
virtual architecture so far should be a Guggenheim
Virtual Museum, an institution notable for pioneering
organic, flowing spaces in its real buildings.

**Guggenheim
Virtual Museum**
2000–2001
New York City,
New York, USA

Designed to
provide global
internet access to
the Guggenheim
Museum collec-
tions, as well as
its own digital and
internet art and
the usual museum
amenities, the
Virtual Museum
seems to fit neatly
in the Guggen-
heim's catalogue
of real buildings
that challenge
the norms of
architectural
space planning

Jakob + MacFarlane

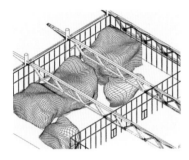

Restaurant Georges
Centre Pompidou, Paris, France
1998–2000

Computer graphics show how the virtual grid of the intervention surfaces bubble up from the structural grid of the Pompidou Centre floor and wrap in an ad hoc fashion around the kitchen and other service spaces

At first sight, the Restaurant Georges on the sixth floor of Piano and Rogers's Centre Pompidou seems to provide the archetypal example of an organic form placed in dialogue with a machine void. Yet Dominique Jakob and Brendan MacFarlane's intervention in this famous totem of modern architecture turns out to share the same intention of using minimum materiality to provide maximum flexibility.

Jakob and MacFarlane felt the onus of responding to 'such a particular architectural context'. Rather than insert a massive architectural contrast (as Gae Aulenti did at the Musée d'Orsay, for example), they sought to make 'the lightest possible intervention'. The floor provided the starting point, reimagined as a deformable surface beneath and behind which the spaces needed to service the restaurant could be hidden. The fragmented, distorted surface was realized in brushed aluminium to diffuse reflected light, achieving 'a minimal presence with a strong personality'.

The various volumes injected, as it were, below this skin contain the kitchen and ancillary spaces for the restaurant. These spaces were planned individually according to normal functionalist criteria. The skin was then stretched in the computer model to accommodate the form and position of these areas.

Designs of this kind typically develop through a potentially infinite set of iterations or 'evolutions' worked out by computer, but in practice a halt must be called at some point determined by human intervention. The architects wished to commemorate this moment by freezing the project just when it began to seem unsupportable, where convexities seemed about to burst and other areas apparently lost turgidity and became flaccid (to use two technical terms from biology that the architects found helpful). 'We stopped where we thought it was neither one thing nor the other to give the spectator the impression that there's an event that's unfolding,' says MacFarlane.

Restaurant Georges

Sponge-like aluminium blobs appear to grow out of the floor, defining spaces in the Georges restaurant on the sixth floor of the Centre Pompidou. 'This "real" grid we appropriated as our "conceptual" one, which then becomes deformed by the volumes or pockets,' explains MacFarlane. The softly reflective, fluid forms of Jakob + MacFarlane's restaurant contrast effectively with the machine aesthetic of the host building

Loewy Bookshop
Paris, France
2001

The square grid of shelves in the Loewy bookshop provides an effective foil for the organic form of the floor-to-ceiling shelf units. The design may amuse bibliophiles by exposing the spurious sense of compendious order implied by standard shelves, but it is a reminder to architects that fluid form is often perversely dependent on Cartesian structural modules

envoi: biomimesis

A generation ago, Philip Steadman, who has otherwise been busy making an extended case for biological analogy in architecture, could write that 'when all is said and done, the fact is that buildings, machines and implements are inert physical objects and not organisms; and the relevance of biological ideas to their study can only remain in the end of an analogical and metaphorical nature.'

If this was true in 1979 when Steadman's *The Evolution of Designs* was published, it's certainly not true now. There is today a widespread interest in buildings that respond actively to their environment, which indicates a deeper relevance of biological similitude to their function and behaviour as well as to their appearance.

It is historically axiomatic that, while environments may alter, buildings tend to stay the same. Indeed, monumentalism has been a defining characteristic of the art of architecture. Living organisms, meanwhile, are both adapted to their environment over the long term by evolution, and capable of responding in various ways to its changes from moment to moment and day to day.

Similar adaptation is one of the main goals of those pioneering the emerging field of 'biomimetics'. We would like the 'intelligent' building of a future generation to open its windows like eyelids to the dawn and to sense the heat in the rising sun or respond to the chill of a breeze by raising the hairs on its back for insulation. Whether it does such things literally or metaphorically is now the issue. Of course, it is possible to engineer solutions not unlike this in the old brute fashion, but since nature does these things so well it seems foolish not to sneak a look at her answers first.

One of its pioneers, zoologist-turned-engineer Julian Vincent, defines biomimetics as 'the abstraction of good design from nature'. The qualifier 'good' is important, as is the term 'abstraction' – biomimetics is not about slavish imitation of nature at any cost, but the judicious selection of observed properties and their subsequent development into sophisticated artificial technologies.

The attraction of biomimetics for architects is that it raises the prospect of closer integration of form and function (in this light, biomimetic architecture is seen as an extension of Modernism). It promises to yield new means by which buildings may respond to, and interact with, their users – means more subtle and more satisfying than present mechanical systems. At a deeper level, according to George Jeronimidis of the Centre for Biomimetics at the University of Reading, architects are drawn to the field 'because we are all part of the same biology'. The urge to build in closer sympathy with nature is, he believes, a genuinely biological, and not merely a Romantic, urge.

There are many areas of promise. Animals offer lessons in thermal performance; plants provide models of response to solar radiation; both suggest novel methods of weather protection. The presumption is always that nature's solutions to problems of adaptation, reached after millions of years of evolution, are better – if not in type, then in degree – than the ones achieved by architects and engineers over a few centuries. We may gasp at the latest suspension bridge, but its steel cables still fall far short of their theoretical limit of performance compared, for example, with the silken fibres of a spider's web.

Some recent research studies illustrate the range of phenomena in nature that are inspiring efforts at emulation. Intrigued by the ability of penguins to withstand the cold despite fasting for months while incubating their eggs, a team at the Centre for Biomimetics measured the shape of their feathers in order to construct a thermal model that would tell them about the insulating properties of the bird's coat. This new understanding may lead to improved insulation in buildings or clothing.

Following the observation that the leaves of the lotus plant would appear clean almost as soon as it began to rain, a scientist at Bonn University found that they were covered with waxy hairs that ensure that impacting raindrops do not wet the leaf surface. Instead, the droplets are supported by the hairs and quickly absorb dirt particles caught there, before rolling off the leaf leaving it perfectly clean. This effect has now been replicated in self-cleaning paint.

The valuable lessons to be learnt are not always the most direct ones or the most 'true to nature'. An investigation into why down feathers provide more efficient thermal insulation on birds than they do in down jackets worn by humans, for example, found that the action of the feathers was more important than the feathers themselves, suggesting that a more efficient manmade insulator should look more closely at replicating the mechanism but use superior artificial materials in place of real feathers.

For Singapore's new arts centre, engineers Atelier One looked at the function of polar bear fur. The way that tiny muscles contract to raise the hair and provide an insulating layer of air to protect the skin from the cold was emulated in rudimentary fashion using movable aluminium panels. However, the aim on this occasion was to keep the heat out rather than in, as well as to manage solar gain. The stimulus and response were sunlight and photocell sensors rather than temperature and nerve endings, but nature's mechanism informed the design solution nevertheless.

On a higher plane of ambition, nature does things we would very much like to be able to do in building – for example, clothing an animal in a contiguous layer of skin that is nevertheless variable in its flexibility, toughness and sensitivity to heat, light and touch. Nature also has a pleasing way of organizing structures that possess similar degrees of complexity at different scales. The cellular and fibrous structures of bone, muscle and wood have this quality, giving them a range of useful properties, whereas in manmade materials such as steel and many plastics the complexity is concentrated at one scale.

These inspirations will undoubtedly lead to new materials and technologies. But the pace of development may be slow. The building industry is notoriously conservative, especially when it comes to new materials. It seems readier, surprisingly, to embrace the formal extravagances of its architects than it does the substantial innovations of its engineers and scientists.

Ultimately, there is the wish to learn not just from the forms of life but from the cycle of life. The building-as-organism is a seductive vision. Living organisms are sparing with materials to a degree that a human engineer can only admire, for example, and can afford to be so in part because they are capable of healing their wounds. Unlike animals, buildings can't presently renew or repair even their skin, but they would clearly gain from being able to do so. From self-repair, it is a short conceptual step to the idea of self-assembly, of designs that grow or extend themselves in adaptation to their environment. And from there it is but another step to self-replication.

Implicit, but so far unstated, in this quest is the matter of death. Species evolve, individual organisms develop and reproduce, respond and adapt to their environment – and then they die. Perhaps only when buildings too are able to curl up and die will the biomimetic project be complete, and architecture's age of monumentalism be truly over.

references and select bibliography

Alexander, Christopher, Hajo Neis, Artemis Anninou, and Ingrid King, **A New Theory of Urban Design** (Oxford: Oxford University Press, 1987).

Ball, Philip, **The Self-Made Tapestry: Pattern Formation in Nature** (Oxford: Oxford University Press, 1999).

BMW Erlebnis- und Ausliefer- ungszentrum/ BMW Event and Delivery Centre (Berlin: Aedes, 2002).

Bonner, John T., **Morphogenesis** (Princeton: Princeton University Press, 1952).

Brayer, Marie-Ange, and Béatrice Simonot, eds, **Archilab's Futurehouse: Radical Experiments in Living Space** (London: Thames & Hudson, 2002).

Burnie, David, ed., **Animal** (London: Dorling-Kindersley, 2001).

Burns, Jim, **Arthropods: New Design Futures** (London: Academy Editions, 1972).

Collins, George R., **Fantastic Architecture** (London: Thames & Hudson, 1980).

Collins, Peter, **Changing Ideals in Modern Architecture, 1750–1950** (London: Faber & Faber, 1965).

Dawkins, Richard, **The Blind Watchmaker** (London: Longman, 1986; Harmondsworth: Penguin, 1991).

—, **Climbing Mount Improbable** (Harmondsworth: Penguin, 1996).

Di Cristina, Giuseppa, ed., **Architecture and Science** (Chichester: Wiley-Academy, 2001).

Forty, Adrian, **Words and Buildings: A Vocabulary of Modern Architecture** (London: Thames & Hudson, 2000).

Frampton, Kenneth, **Modern Architecture: A Critical History** (London: Thames & Hudson, 1980).

Frazer, John, **An Evolutionary Architecture** (London: Architectural Association, 1995).

Frisch, Karl von, **Animal Architecture** (New York: Harcourt Brace Jovanovich, 1974).

Galison, Peter, and Emily Thompson, eds, **The Architecture of Science** (Cambridge, Mass.: MIT Press, 1999).

Gordon, Jim E., **Structures, or Why Things Don't Fall Down** (Harmondsworth: Penguin, 1978).

Guillery, Peter, **The Buildings of London Zoo** (London: Royal Commission on the Historical Monuments of England, 1993).

Haeckel, Ernst, **Kunstformen der Natur** (Leipzig/Vienna: Bibliographisches Institut, 1904).

Hagan, Susannah, **Taking Shape: A New Contract between Architecture and Nature** (Oxford: Architectural Press, 2001).

Hansell, Mike, **Animal Architecture and Building Behaviour** (London: Longman, 1984).

—, **The Animal Construction Company** (Glasgow: Hunterian Museum, 1999).

Jencks, Charles, **Architecture 2000** (London: Studio Vista, 1971).

—, **The Language of Post-Modern Architecture**, 4th edn (London: Academy Editions, 1984).

—, **The Architecture of the Jumping Universe** (London: Academy Editions, 1995).

—, **Architecture 2000 and Beyond: Success in the Art of Prediction** (Chichester: Wiley-Academy, 2000).

—, **The New Paradigm in Architecture: The Language of Post-Modernism** (New Haven: Yale University Press, 2002).

Jung, Carl, **Man and his Symbols** (London: Aldus, 1964; Basingstoke: Picador, 1978).

Kelly, Kevin, **Out of Control: The New Biology of Machines** (London: Fourth Estate, 1994).

Kiesler, Frederick J., **Selected Writings** (Ostfildern bei Stuttgart: Gerd Hatje, 1996).

Le Corbusier, **Towards a New Architecture**, trans. Frederick Etchells (London: Architectural Press, 1927; 1946).

Lethaby, William R., **Form in Civilization** (London: Oxford University Press, 1922).

—, **Architecture, Nature and Magic** (London: Duckworth, 1956).

Lloyd Jones, David, **Architecture and the Environment: Bioclimatic Building Design** (London: Laurence King, 1998).

Lupton, Ellen, **Skin** (New York: Cooper-Hewitt Museum, 2002).

Lynn, Greg, **Folds, Bodies and Blobs: Collected Essays** (Brussels: La Lettre Volée, 1998).

—, **Animate Form** (New York: Princeton Architectural Press, 1999).

Margulis, Lynn, and Karlene Schwartz, **Five Kingdoms: An Illustrated Guide to the Phyla of Life on Earth** (San Francisco: W.H. Freeman, 1982).

Migayrou, Frédéric, and Marie-Ange Brayer, **Archilab: Radical Experiments in Global Architecture** (London: Thames & Hudson, 2001).

Otto, Frei, ed., **Tensile Structures**, 2 vols (Cambridge, Mass.: MIT Press, 1967).

Pearson, David, **New Organic Architecture: The Breaking Wave** (London: Gaia Books, 2001).

Pérez-Gómez, Alberto, **Architecture and the Crisis of Modern Science** (Cambridge, Mass.: MIT Press, 1993).

Portoghesi, Paolo, **Nature and Architecture**, trans. Erika G. Young (Milan: Skira, 2000).

Powers, Alan, **Nature in Design** (London: Conran Octopus, 1999).

Purves, William, David Sadava, Gordon Orians, and H. Craig Heller, **Life: The Science of Biology**, 6th edn (Sunderland, Mass.: Sinauer Associates, 2001).

Ragheb, J. Fiona, **Frank Gehry, Architect** (New York: Guggenheim Museum, 2001).

Rashid, Hani, and Lise-Anne Couture, **Flux** (London: Phaidon, 2002).

Rowe, Colin, **The Mathematics of the Ideal Villa** (Cambridge, Mass.: MIT Press, 1976).

Rudofsky, Bernard, **Architecture without Architects** (New York: Museum of Modern Art, 1964).

Selz, Peter, and Mildred Constantine, eds, **Art Nouveau: Art and Design at the Turn of the Century** (New York: Museum of Modern Art, 1959).

Semper, Gottfried, **The Four Elements of Architecture and Other Writings**, trans. H.F. Mallgrave and W. Herrmann (Cambridge: Cambridge University Press, 1989).

Sorkin, Michael, **Michael Sorkin Studio: Wiggle** (New York: Monacelli Press, 1998).

Steadman, Philip, **The Evolution of Designs: Biological Analogy in Architecture and the Applied Arts** (Cambridge: Cambridge University Press, 1979).

Steele, James, **Architecture and Computers** (London: Laurence King, 2001).

Thompson, D'Arcy W., **On Growth and Form** (Cambridge: Cambridge University Press, 1917; abridged edn 1961, Canto edn 1992).

Topham, Sean, **Blowup: Inflatable Art, Architecture and Design** (Munich: Prestel, 2002).

Tsui, Eugene, **Evolutionary Architecture: Nature as a Basis for Design** (New York: Wiley, 1999).

Venturi, Robert, Denise Scott Brown, and Steven Izenour, **Learning from Las Vegas** (Cambridge, Mass.: MIT Press, 1972).

Weston, Richard, **Utzon** (Hellerup: Edition Bløndal, 2002).

Weyl, Hermann, **Symmetry** (Princeton: Princeton University Press, 1952; 1989).

Wilkinson, Chris, and Jim Eyre, **Bridging Art and Science** (London: Booth-Clibborn, 2001).

picture credits

The author and Laurence King Publishing Limited thank the sources of photographs for supplying and granting permission to reproduce them. Photographs have been obtained from the architects responsible unless listed below. Every effort has been made to contact all copyright holders. Laurence King Publishing Limited would be pleased to hear from any such source that could not be traced.

10 AKG London; **11** courtesy Ford Motor Company; **12** Private collection; **13** Bridgeman Art Library; **15L** Bastin & Evrard; **15R** A.F. Kersting, London; **16** Tamás Szántó; **17** Ralph Pugliese Jr. © 2000; **18** © Donald Corner & Jenny Young/ GreatBuildings.com; **19** RIBA; **20** AKG London; **24** Private collection; **26** Christian Richters; **32L** courtesy Times Books; **32TC** A.F. Kersting, London; **32BC** Alinari; **32TR** courtesy Parco dei Mostri, Bomarzo; **32BR** courtesy Vor Frelsers Kirke; **33L** Private collection; **34L** photo courtesy of The Save Lucy Committee, Inc. ©2002 All Rights Reserved; **35L** AKG London; **35C** photo by Donna Coveney, MIT; **35R** Paul M. R. Maeyaert; **36L** courtesy Richard Weston; **37TC** Ronald Grant Archive; **37BC** AKG London; **38–39** Birds Portchmouth Russum; **46** Hisao Suzuki; **47** © Michael Webb; **49** © Roland Halbe; **70, 71L** Nicholas Kane; **80–81, 91B** Benedict Luxmoore; **92, 93** © Moreno Maggi; **100, 101** Richard Davies; **102, 103** © Matteo Piazza; **111** © Keegan/ Duigenan; **112–113** Festo; **86B** © Melon; **114–115** © Read & Peck; **116B** © Keith Collie; **117L** © Peter Strobel; **118** © Peter Cook; **119** © Richard Kalina; **120BL** © Herbie Knott; **120BR** © Edmund Sumner; **121T** © Photodesign/ Jens Willebrand; **121B** © Archtekton/ Werner Huthmacher; **125M** © Judit Kimpian; **130-131** NOX; **167T & BR** Nicolas Borel.

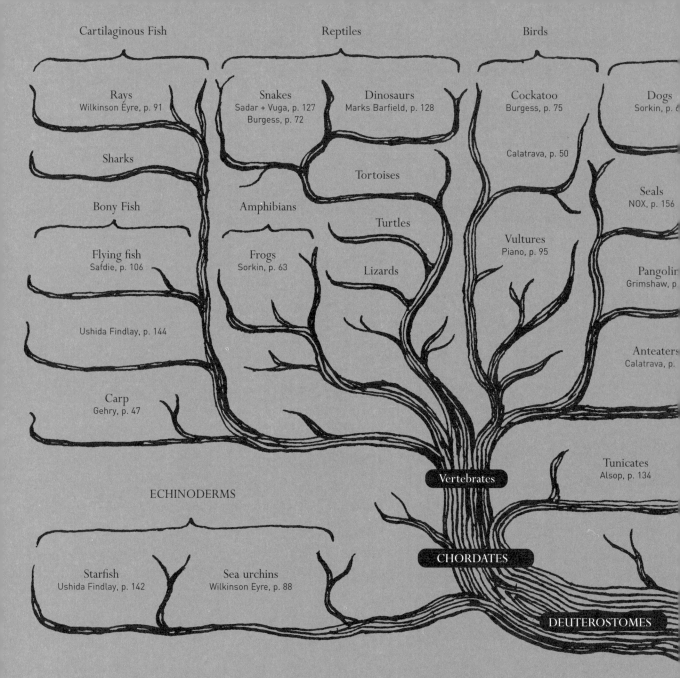

Cartilaginous Fish

Rays
Wilkinson Eyre, p. 91

Sharks

Bony Fish

Flying fish
Safdie, p. 106

Ushida Findlay, p. 144

Carp
Gehry, p. 47

Reptiles

Snakes
Sadar + Vuga, p. 127
Burgess, p. 72

Dinosaurs
Marks Barfield, p. 128

Tortoises

Turtles

Amphibians

Frogs
Sorkin, p. 63

Lizards

Birds

Cockatoo
Burgess, p. 75

Calatrava, p. 50

Vultures
Piano, p. 95

Dogs
Sorkin, p. 6

Seals
NOX, p. 156

Pangolin
Grimshaw, p.

Anteaters
Calatrava, p.

Tunicates
Alsop, p. 134

Vertebrates

ECHINODERMS

CHORDATES

Starfish
Ushida Findlay, p. 142

Sea urchins
Wilkinson Eyre, p. 88

DEUTEROSTOMES

Key to animal classification:

UPPER CASE: **phylum or higher**
lower case: **class or lower**
Sources: **Burnie**, *Animal*; **Margulis and Schwartz**, *Five Kingdoms*; http://tolweb.org